Linkages Between Arctic Warming and Mid-Latitude Weather Patterns

Summary of a Workshop

Katie Thomas, Rapporteur

Committee on Linkages Between Arctic Sea Ice Loss
and Mid-Latitude Weather Patterns: A Workshop

Board on Atmospheric Sciences and Climate

Polar Research Board

Division on Earth and Life Studies

NATIONAL RESEARCH COUNCIL
OF THE NATIONAL ACADEMIES

THE NATIONAL ACADEMIES PRESS
Washington, D.C.
www.nap.edu

THE NATIONAL ACADEMIES PRESS 500 Fifth Street, NW Washington, DC 20001

NOTICE: The project that is the subject of this report was approved by the Governing Board of the National Research Council, whose members are drawn from the councils of the National Academy of Sciences, the National Academy of Engineering, and the Institute of Medicine. The members of the committee responsible for the report were chosen for their special competences and with regard for appropriate balance.

This study was supported by the National Science Foundation under contract numbers AGS-1158602 and ARC-1137066, the National Aeronautics and Space Administration under contract numbers NNX08AB07G and NNX10AR69G, and the National Oceanic and Atmospheric Administration under contract numbers NA11OAR4600211 and NA13OAR4310215. Any opinions, findings, and conclusions, or recommendations expressed in this material are those of the author(s) and do not necessarily reflect the views of the sponsoring agencies or any of their subagencies.

International Standard Book Number-13: 978-0-309-30188-6
International Standard Book Number-10: 0-309-30188-2

Additional copies of this report are available for sale from the National Academies Press, 500 Fifth Street, NW, Keck 360, Washington, DC 20001; (800) 624-6242 or (202) 334-3313; http://www.nap.edu/ .

Printed in the United States of America

THE NATIONAL ACADEMIES

Advisers to the Nation on Science, Engineering, and Medicine

The **National Academy of Sciences** is a private, nonprofit, self-perpetuating society of distinguished scholars engaged in scientific and engineering research, dedicated to the furtherance of science and technology and to their use for the general welfare. Upon the authority of the charter granted to it by the Congress in 1863, the Academy has a mandate that requires it to advise the federal government on scientific and technical matters. Dr. Ralph J. Cicerone is president of the National Academy of Sciences.

The **National Academy of Engineering** was established in 1964, under the charter of the National Academy of Sciences, as a parallel organization of outstanding engineers. It is autonomous in its administration and in the selection of its members, sharing with the National Academy of Sciences the responsibility for advising the federal government. The National Academy of Engineering also sponsors engineering programs aimed at meeting national needs, encourages education and research, and recognizes the superior achievements of engineers. Dr. C. D. Mote, Jr., is president of the National Academy of Engineering.

The **Institute of Medicine** was established in 1970 by the National Academy of Sciences to secure the services of eminent members of appropriate professions in the examination of policy matters pertaining to the health of the public. The Institute acts under the responsibility given to the National Academy of Sciences by its congressional charter to be an adviser to the federal government and, upon its own initiative, to identify issues of medical care, research, and education. Dr. Harvey V. Fineberg is president of the Institute of Medicine.

The **National Research Council** was organized by the National Academy of Sciences in 1916 to associate the broad community of science and technology with the Academy's purposes of furthering knowledge and advising the federal government. Functioning in accordance with general policies determined by the Academy, the Council has become the principal operating agency of both the National Academy of Sciences and the National Academy of Engineering in providing services to the government, the public, and the scientific and engineering communities. The Council is administered jointly by both Academies and the Institute of Medicine. Dr. Ralph J. Cicerone and Dr. C. D. Mote, Jr., are chair and vice chair, respectively, of the National Research Council.

www.national-academies.org

COMMITTEE ON LINKAGES BETWEEN ARCTIC SEA ICE LOSS AND MID-LATITUDE WEATHER PATTERNS: A WORKSHOP

DAVID A. ROBINSON, (*Chair*), Rutgers, The State University of New Jersey, Piscataway
UMA BHATT, University of Alaska, Fairbanks
CECILIA BITZ, University of Washington, Seattle
LANCE F. BOSART, State University of New York, Albany
DAVID H. BROMWICH, Ohio State University, Columbus
CLARA DESER, National Center for Atmospheric Research, Boulder, CO
WALTER MEIER, University of Colorado, Boulder

NRC Staff:

KATIE THOMAS, Study Director
AMANDA STAUDT, Director
SUSAN ROBERTS, Director (until September 2013)
LAURIE GELLER, Senior Program Officer
AMANDA PURCELL, Research and Financial Associate
RITA GASKINS, Administrative Coordinator
ROB GREENWAY, Program Associate

BOARD ON ATMOSPHERIC SCIENCES AND CLIMATE

ANTONIO J. BUSALACCHI, JR. (*Chair*), University of Maryland, College Park
GERALD A. MEEHL (*Vice Chair*), National Center for Atmospheric Research, Boulder, Colorado
LANCE F. BOSART, State University of New York, Albany
MARK A. CANE, Columbia University, Palisades, New York
SHUYI S. CHEN, University of Miami, Florida
HEIDI CULLEN, Climate Central, Princeton, New Jersey
PAMELA EMCH, Northrup Grumman Aerospace Systems, Redondo Beach, California
WILLIAM B. GAIL, Global Weather Corporation, Boulder, Colorado
LISA GODDARD, Columbia University, Palisades, New York
TERRI S. HOGUE, Colorado School of Mines, Golden
ANTHONY JANETOS, Joint Global Change Research Institute, College Park, Maryland
RONALD "NICK" KEENER, JR., Duke Energy Corporation, Charlotte, North Carolina
JOHN E. KUTZBACH, University of Wisconsin-Madison
STEPHEN W. PACALA, Princeton University, New Jersey
ARISTIDES A.N. PATRINOS, New York University, Brooklyn
RAYMOND T. PIERREHUMBERT, The University of Chicago, Illinois
KIMBERLY PRATHER, University of California, San Diego
S.T. RAO, North Carolina State University, Raleigh
DAVID A. ROBINSON, Rutgers, The State University of New Jersey, Piscataway
JOHN T. SNOW, The University of Oklahoma, Norman
CLAUDIA TEBALDI, Climate Central, Princeton, New Jersey
XUBIN ZENG, University of Arizona, Tucson

Ocean Studies Board Liaison

DANIEL RUDNICK, University of California, San Diego

NRC Staff

AMANDA STAUDT, Director
EDWARD DUNLEA, Senior Program Officer
LAURIE GELLER, Senior Program Officer
KATHERINE THOMAS, Program Officer
LAUREN BROWN, Associate Program Officer
AMANDA PURCELL, Research and Financial Associate
RITA GASKINS, Administrative Coordinator
ROB GREENWAY, Program Associate
SHELLY FREELAND, Senior Program Assistant

POLAR RESEARCH BOARD

JAMES W. C. WHITE (*Chair*), University of Colorado, Boulder
WALEED ABDALATI, University of Colorado, Boulder
SRIDHAR ANANDAKRISHNAN, Pennsylvania State University, University Park
KATEY WALTER ANTHONY, University of Alaska, Fairbanks
JULIE BRIGHAM-GRETTE, University of Massachusetts, Amherst
JOHN CASSANO, University of Colorado, Boulder
JENNIFER A. FRANCIS, Rutgers University, Marion, Massachusetts
EILEEN E. HOFMANN, Old Dominion University, Norfolk, Virginia
BERNICE M. JOSEPH, University of Alaska, Fairbanks (until January 13, 2014)
ELLEN S. MOSLEY-THOMPSON, Ohio State University, Columbus
GEORGE B. NEWTON, U.S. Arctic Research Commission, Marstons Mills, Massachusetts
RAFE POMERANCE, Independent Consultant, Washington, DC
CARYN REA, ConocoPhillips, Anchorage, Alaska
GAIUS R. SHAVER, The Ecosystems Center, Woods Hole, Massachusetts
ALLAN T. WEATHERWAX, Siena College, Loudonville, New York

Ex-Officio

JACQUELINE M. GREBMEIER (*U.S. Delegate to IASC*), University of Maryland, Solomons
TERRY WILSON (*U.S. Delegate to SCAR*), Ohio State University, Columbus
DENEB KARENTZ (*Alternate U.S. Delegate to SCAR*), University of San Francisco, California

NRC Staff

AMANDA STAUDT, Director
EDWARD DUNLEA, Senior Program Officer
LAURIE GELLER, Senior Program Officer
KATHERINE THOMAS, Program Officer
LAUREN BROWN, Associate Program Officer
AMANDA PURCELL, Research and Financial Associate
RITA GASKINS, Administrative Coordinator
ROB GREENWAY, Program Associate
SHELLY FREELAND, Senior Program Assistant

Acknowledgments

This report has been reviewed in draft form by individuals chosen for their diverse perspectives and technical expertise, in accordance with procedures approved by the National Research Council's (NRC's) Report Review Committee. The purpose of this independent review is to provide candid and critical comments that will assist the institution in making its published report as sound as possible and to ensure that the report meets institutional standards for objectivity, evidence, and responsiveness to the study charge. The review comments and draft manuscript remain confidential to protect the integrity of the deliberative process. The committee wishes to thank the following individuals for their review of this report:

CECILIA BITZ, University of Washington, Seattle
ISAAC HELD, National Oceanic and Atmospheric Administration, Princeton, NJ
MARK SERREZE, University of Colorado, Boulder
JOHN WALSH, University of Alaska, Fairbanks
MARGARET LEMONE, National Center for Atmospheric Research, Boulder, CO

Although the reviewers listed above have provided constructive comments and suggestions, they were not asked to endorse the views of the committee, nor did they see the final draft of the report before its release. The review of this report was overseen by **Robin Bell,** Columbia University, Palisades, NY; appointed by the NRC, she was responsible for making certain that an independent examination of this report was carried out in accordance with institutional procedures and that all review comments were carefully considered. Responsibility for the final content of this report rests entirely with the authoring panel and the institution.

Contents

Linkages Between Arctic Warming and Mid-Latitude Weather Patterns

OVERVIEW

The Arctic has been undergoing significant changes in recent years. Surface temperatures in the region are rising twice as fast as the global mean. The extent and thickness of sea ice is rapidly declining. Such changes may have an impact on atmospheric conditions outside the region. Several hypotheses for how Arctic warming may be influencing mid-latitude weather patterns have been proposed recently. For example, Arctic amplified warming could lead to a weakened jet stream resulting in more persistent weather patterns in the mid-latitudes. Or Arctic sea ice loss could lead to an increase of snow on high-latitude land; snow expands on land in autumn, which in turn impacts the jet stream resulting in cold Eurasian and North American winters. These and other potential connections between a warming Arctic and mid-latitude weather are the subject of active research.

The National Research Council convened a workshop on September 12-13, 2013, to review our current understanding and to discuss research needed to better understand proposed linkages. The workshop participants were encouraged to take a global perspective and consider the influence of the Arctic in the context of forcing from other components of the climate system, such as changes in the tropics, ocean circulation, and mid-latitude sea surface temperature.

Many workshop participants noted that research on these linkages is still in its infancy, making it difficult to draw conclusions regarding their existence or their mechanisms. Workshop presentations highlighted the complex mechanisms and feedbacks that link the tropics, poles, and mid-latitudes while also noting that they can be bidirectional and not necessarily direct (e.g., the stratosphere and oceans may act as intermediaries). Understanding how Arctic warming could impact these linkages or other large-scale circulation patterns is critical, but the large natural variability of the atmosphere and Arctic sea ice makes attribution difficult.

Furthermore, many of the proposed linkages are composed of multistep hypotheses, and our understanding of the various steps is complicated by inconsistent evidence from observations and modeling studies. Some participants said that the time series for observations is too short to detect statistically significant trends. It was noted by some participants that trends could be clarified through the quantification of uncertainty in data products. Atmospheric reanalyses and efforts to extend Arctic sea ice records back in time might also be useful for detecting trends. Under-utilized observations could also be harnessed from often overlooked sources, such as the National Ice Center (NIC) operational products.

There is also no standard for objective measurement of important mechanisms such as wave amplitude or atmospheric blocks, which lead to a stagnation of weather patterns. A better understanding of the range of blocking systems and how they are impacted by various components of the climate system is needed. Participants suggested deploying process-based studies on the "weak links" of the specific hypotheses as a way forward.

Results from modeling studies offer a lot of divergence and are difficult to compare systematically because different boundary and/or initial conditions are often used. To address this challenge, many participants would like the modeling community to organize careful model intercomparisons and conduct model sensitivity studies with identical initial and/or boundary conditions. Moreover, given that sea ice loss is only one driver of Arctic amplification, many participants said attribution studies are needed to understand the proportion of Arctic amplification that is due to processes outside of the Arctic. Although models are useful for providing clues on the mechanisms of potential Arctic linkages, as well as detection and attribution, some participants acknowledged that there are important biases and limitations that must be considered.

Many participants said that combining observations with a hierarchy of models is key to making progress on this issue. Furthermore, attention so far has focused on changes in the meridional temperature gradient, but future studies might also consider other mechanisms of Arctic linkages, such as zonal temperature gradients, static stability, and moisture changes. The climate and weather communities could collaborate to apply the sophisticated methods and diagnostics developed in the meteorological community to better understand linkages. Some participants also thought that the possibility of regional impacts of Arctic warming is an important consideration.

Some workshop participants noted that a large-scale research program dedicated to understanding the mechanisms that link a warming Arctic and the heavily-populated mid-latitudes is needed, particularly because it could lead to improved seasonal forecasts. Several participants pointed to the analogous example of an improved understanding of El Niño Southern Oscillation (ENSO) resulting in improved seasonal forecast skill. Participants expressed optimism that significant progress could be obtained because of several recent achievements and events: (a) improvements in atmospheric, oceanic, and sea ice models; (b) longer observational records of sea ice and atmospheric circulation; (c) emergence of a reasonably strong sea ice forcing in recent years; and (d) strong interest in the topic from policymakers, scientists, and the public.

INTRODUCTION

On September 12-13, 2013 the Board on Atmospheric Sciences and Climate and the Polar Research Board held a public workshop that brought together a diverse array of experts to examine linkages between a warming Arctic and mid-latitude weather patterns. The workshop included presentations from leading researchers representing a range of views on this topic.

This workshop was planned and organized by an ad hoc committee of the National Research Council[1]. The committee was tasked to plan a workshop to address the following questions:

1. What do we currently understand about the mechanisms that link declines in Arctic sea ice cover, loss of high-latitude snow cover, changes in Arctic-region

[1] This report has been prepared by the workshop rapporteur as a summary of what occurred at the workshop. The planning committee's role was limited to planning and convening the workshop. The views contained in the report are an interpretation of those presented by individual workshop participants and do not necessarily represent the views of all workshop participants, the planning committee, or the National Research Council.

energy fluxes, atmospheric circulation patterns, and the occurrence of extreme weather events?

2. What may be the possible implications of more severe loss (and eventually, total loss) of summer Arctic sea ice upon weather patterns at lower latitudes?
3. What are the major gaps in our understanding, and what sort of observational and/or modeling efforts are needed to fill those gaps?
4. What are the current opportunities and limitations for using Arctic sea ice predictions to assess the risk of temperature/precipitation anomalies and extreme weather events over northern continents? How might these capabilities improve over time?

The workshop discussions focused on questions 1 and 3. As the research matures, it may become possible to fully address questions 2 and 4 in the future. Committee members planned the workshop structure, identified speakers and attendees, and developed background materials for attendees. They also led the workshop and served as session facilitators. Furthermore, during the workshop planning, they decided that the workshop should expand to include links to Arctic amplification as well, because Arctic amplification is not exclusively driven by sea ice loss.

All workshop presentations are available online at http://dels.nas.edu/global/basc/al-presentations. This summary was written by a rapporteur to present the various ideas and suggestions that arose in the workshop discussion and to synthesize the main discussion items. It does not include any conclusions or recommendations, nor does it cover the full spectrum of issues around this topic. For further information, relevant references that were suggested by workshop participants and planning committee members can be found at http://dels.nas.edu/resources/static-assets/basc/miscellaneous/basc-arctic-linkages-workshopreferences.pdf.

Report Roadmap

The structure of this report largely follows the structure of the workshop sessions and is divided into three general topics: the big picture context, observations, and modeling and theoretical work. Some speakers touched on all three topics in their remarks so this summary does not step through each presentation in detail, but rather gathers related points thematically. Brief abstracts of each presentation are in Appendix A. The final section lists research needs and key messages that were identified by individuals in breakout groups.

ARCTIC WARMING AND EXTREME EVENTS IN THE MID-LATITUDES: A POSSIBLE LINK?

Rising global average temperatures, and especially intense warming in the northern polar regions, are leading to a rapid loss of the sea ice cap that covers the Arctic Ocean. The seven summers with the lowest sea ice minimums all occurred during the past seven years. Workshop presenter James Screen, University of Exeter, noted that Arctic sea ice loss in summer 2012 broke the previous record low set in 2007, with ice cover half as extensive as it was only 30 years ago. Although summer sea ice extent in 2013 did not set a new record low, it still was consistent with a long-term trend of rapid decline. The observed 2012 record low in summer Arctic sea ice extent is extremely unlikely to have occurred due to internal climate variability according to Rong Zhang, National Oceanic and Atmospheric Administration (NOAA).

In addition to sea ice, spring snow cover has also decreased in the Arctic. As discussed by workshop presenter James Overland, NOAA, many areas of the Arctic experienced almost a 40 percent decrease in snow cover in June 2012 compared to the average from 1971-2000 (Figure 1). David Robinson, Rutgers University, noted that annual snow cover extent over

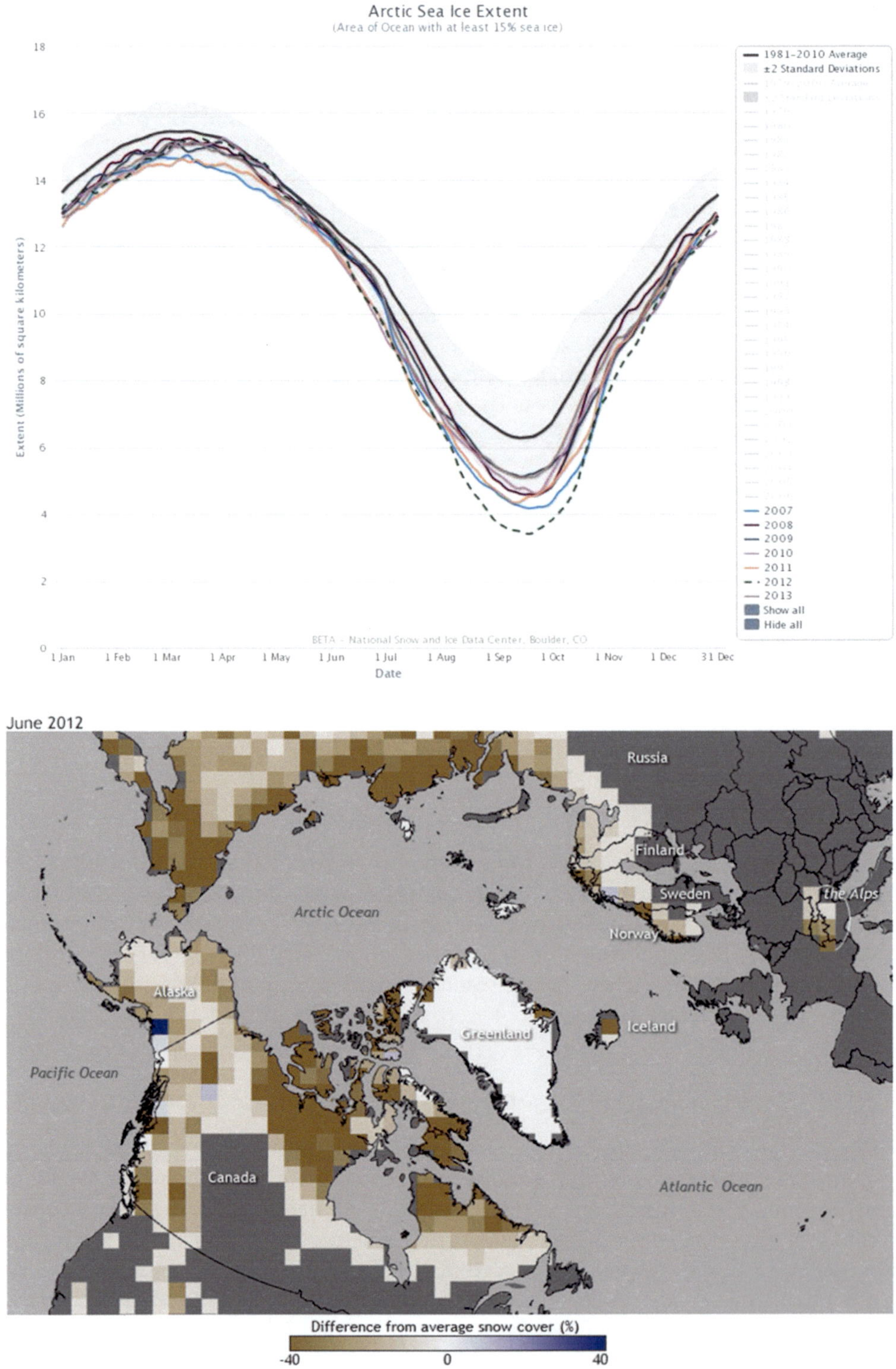

FIGURE 1 Top: Arctic sea ice extent for the years 2007-2013 compared to the 1981-2010 average. Bottom: June snow cover anomalies in June 2012 in the Northern Hemisphere compared to the long-term 1971-2000 average based on the number of days in the year when a location was snow covered. Shades of brown indicate places that experienced up to 40 percent fewer snow-covered days than average in June 2012. Blue indicates areas that experienced up to 40 percent more snow-covered days than average. SOURCE: Climate.gov, http://www.climate.gov/news-features/featured-images/record-low-spring-snow-cover-northern-hemisphere-2012.

Northern Hemisphere lands has averaged lower since the late 1980s than earlier in the satellite era which began in the late 1960s.

The loss of Arctic sea ice has occurred at rates much faster than climate models have recently projected, even for the worst-case forcing scenarios. The full disappearance of late-summer Arctic sea ice is possible in the coming decades (Massonnet et al., 2012; Stroeve et al., 2012). Many people may consider this stark indicator of climate change to be a distant concern that does not affect our "real lives." But such trends do indeed matter greatly to society, because the Arctic may play an integral role in the planetary system of climate and weather well beyond the Arctic region.

Arctic Amplification

Although the workshop topic originally focused on the impact of *Arctic sea ice loss*, because many workshop participants stressed that sea ice loss is only part of the issue, the discussion was broadened to include the impact of *Arctic amplification*. This term is commonly defined as amplified warming near the Arctic compared to the rest of the hemisphere or globe in response to a change in global climate forcing (e.g., the concentration of greenhouse gases [GHGs] or solar output). For example, as the sea ice melts because of warmer temperatures in the Arctic, the resulting darker ocean waters are exposed to incoming solar radiation. The darker ocean absorbs significantly more energy than the white, reflective sea ice. The result is an increase in water and air temperatures, which melts even more sea ice. Overland noted that this heat does not stay in the ocean; it eventually is re-emitted to the atmosphere. The result is a smaller difference in temperature between the frozen north and the more temperate mid-latitudes (an important concept for several of the proposed Arctic linkages). The temperature change associated with Arctic amplification occurs not only near the surface, but also through a considerable depth of the troposphere.

Some participants noted that there are several other drivers of Arctic amplification and that the impact of albedo on Arctic amplification may not be as significant as previously thought (Pithan and Mauritsen, 2014; Winton, 2006). Martin Hoerling, NOAA, suggested that the estimated observed 1000-500hPa (hectopascal) Arctic warming could be largely due to natural decadal sea surface temperature (SST) forcing. See also Feldstein's presentation on the role of the tropics in Arctic Amplification (Appendix A).

Extreme Weather Events

This decade has seen a series of record-breaking extreme weather events taking place around the world (Figure 2). Screen highlighted some severe weather events (e.g., heat waves, drought, flooding) in the mid-latitudes, some of which have been high profile and high impact (e.g., Hurricane Sandy, Russian heat wave). Such events can cause human suffering and have a significant economic impact. In 2011, 14 events in the United States caused losses in excess of US$1 billion each (WMO, 2011).

Proposed Arctic Linkages

The recent extreme weather events and the recent significant decline in Arctic sea ice have spurred an increasing interest in the relationship between a warming Arctic and large-scale climate dynamics, which may have tremendous implications for society. Of particular concern is emerging research that may indicate that Arctic warming can have dramatic impacts upon weather patterns across the heavily populated northern mid-latitudes, and

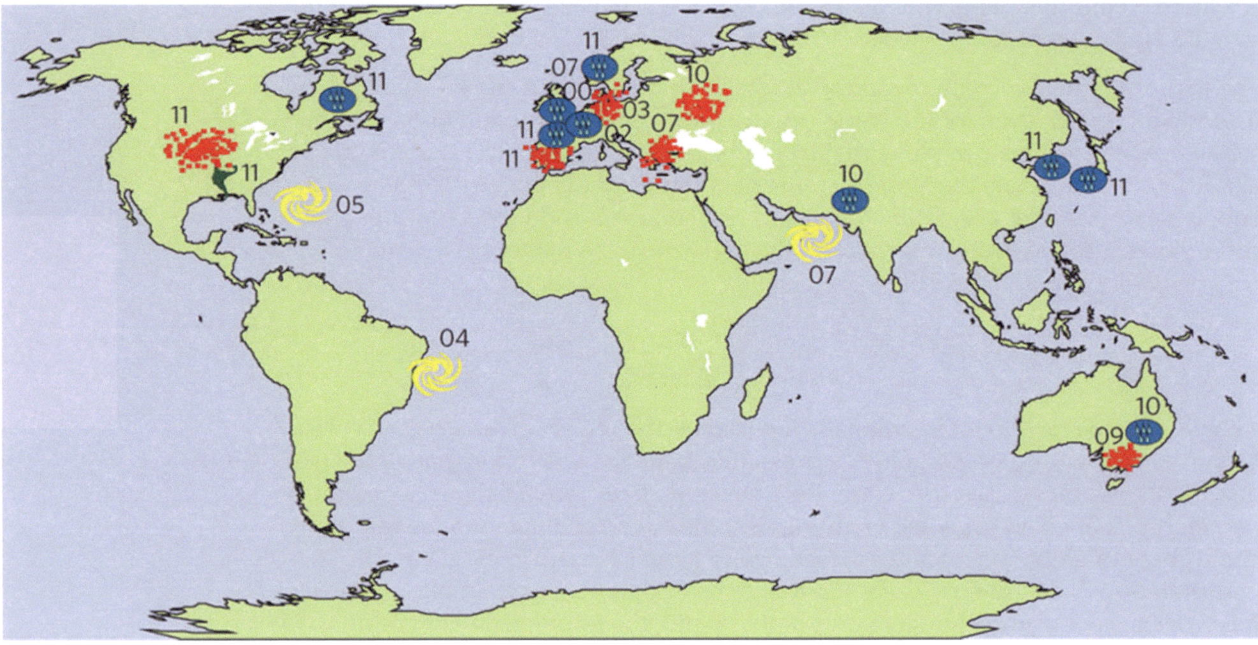

FIGURE 2 World map showing the record-breaking extreme weather events in the past decade. The numbers refer to the year in the twenty-first century. Blue symbols represent rainfall; red symbols represent heat-waves and droughts; yellow symbols represent hurricanes and cyclones; and the green symbol represents a tornado outbreak. SOURCE: Coumou and Rahmstorf, 2012.

that such impacts could increase as ice cover continues to retreat and the Arctic continues to warm in the coming decades (see Box 1).

THE ROLE OF ARCTIC WARMING IN THE CONTEXT OF OTHER FORCING FACTORS

The climate system reflects a complex combination of many interconnected physical and (often chaotic) dynamical processes operating at a variety of timescales in the atmosphere, ocean, and land. A significant portion of the workshop was dedicated to understanding the role of Arctic warming compared to other forcing components of the climate system (e.g., variations in the tropics and stratosphere) in influencing weather in the mid-latitudes. Speakers discussed the various connections and linkages within the climate system and considered whether a warming Arctic now has a stronger influence on atmospheric circulation patterns (e.g., polar vortex, jet stream, modes of variability) and the progression of systems (e.g., blocking).

Large-scale Atmospheric Circulation Patterns: What Drives Mid-Latitude Weather?

Workshop presenter Elizabeth Barnes, Colorado State University, provided an overview of atmospheric dynamics of jet streams, mid-latitude waves, and blocking. She stressed that mid-latitude atmospheric variability is composed of dynamical interactions between circulations of a range of spatial and temporal scales.

BOX 1

Examples of Proposed Arctic Linkages

Below is a list of the proposed linkages that were discussed during the workshop. Further details about some of these linkages can be found later in the report.

Increased Arctic warming → weakened temperature gradient → weakened, more meandering jet stream[a] → more persistent weather patterns in the mid-latitude (Francis and Vavrus, 2012)

Arctic sea ice loss → increase in autumn high latitude snow cover → more expansive and strengthened Siberian high pressure in autumn and winter → increase upward propagation of planetary waves → more sudden stratospheric warmings → weakened polar vortex[b] and weakened, more meandering jet stream (Cohen et al., 2012; Ghatak et al., 2012)

Arctic sea ice loss → changes in regional heat and other energy fluxes → unstable polar vortex → cold polar air moves to the mid-latitudes (Overland and Wang, 2010)

Arctic sea ice loss → amore meandering jet stream and winter atmospheric circulation patterns similar to a negative phase of the winter Arctic Oscillation → frequent episodes of atmospheric "blocking" patterns (Liu et al., 2012).

Arctic sea ice loss → southward shift of the jet stream position over Europe in summer → increased frequency of cloudy, cool, and wet summers over northwest Europe (Screen, 2013)

Arctic sea ice loss → winter atmospheric circulation response resembling the negative phase of the Arctic Oscillation → rainfall extremes in the Mediterranean in winter (Grassi et al., 2013)

Arctic sea ice loss → negative phase of the tripole wind pattern → enhanced winter precipitation and declining winter temperature in East Asia (Wu et al., 2013)

[a] The "typical" jet stream meanders north and south, however it can weaken and slow, allowing the meanders to become larger.

[b] The polar vortex is a large-scale region of air that is contained by a strong west-to-east jet stream that circles the polar region. The polar vortex extends from the tropopause (the dividing line between the stratosphere and troposphere) through the stratosphere and into the mesosphere (above 50 km). Cold temperatures are associated with the air inside the vortex. Definition from http://ozonewatch.gsfc.nasa.gov/facts/vortex_NH.html

---

Barnes focused her presentation on the polar jet stream[2] in the Northern Hemisphere, which flows over the middle to northern latitudes (50-60) of North America, Europe, and Asia and their intervening oceans. The daily jet location pattern of the north Atlantic jet (Figure 3) has three distinct peaks, but it varies significantly over decadal time scales.

Jet streams and other north Atlantic and north Pacific circulations and variability behave differently depending on the season. Barnes noted that jet streams and storm tracks are strongest over the oceans during winter and shift further north in the summer. Jet and storm track strength and location are driven by changes in:

- wave propagation and momentum fluxes (e.g. Chen and Held, 2007; Chen et al., 2008; Kidston and Gerber, 2010);
- meridional temperature gradients (e.g. Haarsma et al., 2013a; Harvey et al., 2013);
- static stability (e.g. Frierson et al., 2007; Lu et al., 2010);
- ocean influences on the meridional overturning circulation (e.g. Chen and Held, 2007; Chen et al., 2008; Woollings et al., 2012);
- eddy phase speed increases (from increasing upper-troposphere/lower stratosphere winds [e.g. Chen and Held, 2007; Chen et al., 2008]);

---

[2] Jet streams are fast flowing, narrow air currents in the near tropopause. The major jet streams flow west to east and their paths typically have a wavy shape.

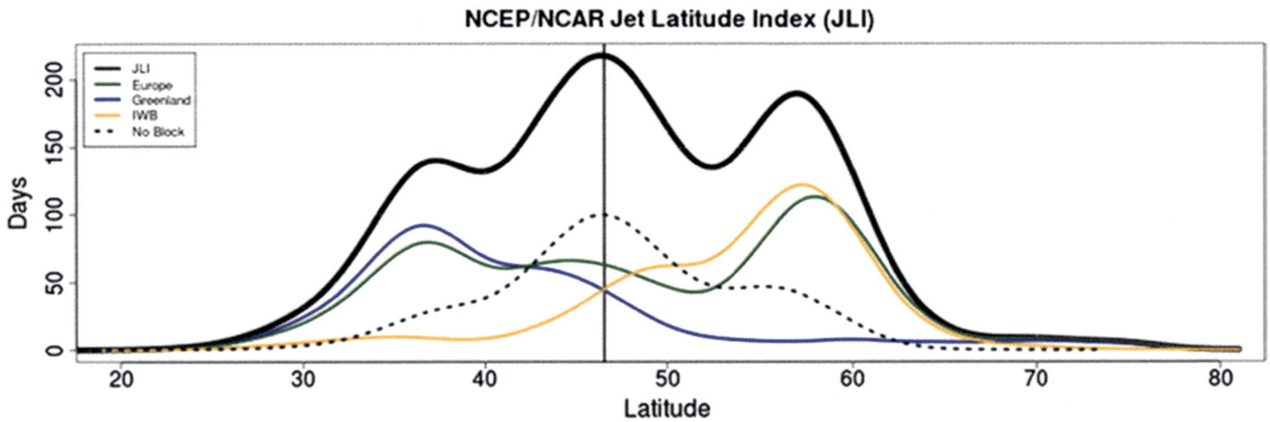

FIGURE 3 Jet latitude index[3] PDF (probability distribution function) for NCEP/NCAR (National Centers for Environmental Prediction/National Center for Atmospheric Research) reanalysis, where the black line represents the climatological distribution; blue line, Greenland blocking days;[4] green line, European blocking days; and orange line, Iberian wave blocking days. The dotted line represents jet latitude index when no blocking in the three sectors is detected. The black vertical line shows the latitude of the central peak. SOURCE: (Davini et al., 2013)

- rising tropopause[5] (e.g. Lorenz and DeWeaver, 2007); and
- changes in wind speed over the region (e.g. Haarsma et al., 2013b; Mizuta, 2012).

Modes of Atmospheric Variability

Modes of variability were discussed at the workshop in the context of linkages (see Box 2 for additional details). The dominant mode of atmospheric variability in the Northern Hemisphere extratropics is the Northern Annular Mode/Arctic Oscillation (NAM/AO), which can influence surface temperature and precipitation, especially the frequency of extreme events. There may be potential for seasonal predictability of the AO (Riddle et al., 2013), but it has been found to be shorter in practice (7-10 days). It is not clear why there seems to be skill in predicting the wintertime AO.

The manifestation of the AO in the Atlantic sector is commonly referred to as the North Atlantic Oscillation (NAO). The NAO is also sometimes referred to as a "Greenland Block." The NAO is the largest contributing pattern to European interannual variability and plays a significant role in predictions of European winter climate. The predictability of the NAO is limited to seasonal timescales (NRC, 2010).

While not extensively discussed at this workshop, participants identified ENSO as an example of a mode of variability in the tropics that can influence weather in the mid-latitudes (i.e., teleconnection). Participants noted that seasonal forecasts improved as our understanding and predictability of ENSO grew; a greater understanding of Arctic influences on mid-latitude weather might allow seasonal (3 months) climate forecasts to be further improved.

---

[3] "Jet latitude index" is defined as the daily latitude of maximum low-level (mass-averaged 925–700 hPa) zonal winds zonally averaged over the Atlantic sector (0 –60 W, 15 –75 N; Woollings et al., 2010).

[4] A Greenland Block is a very strong area of high pressure located over Greenland.

[5] The boundary in the atmosphere between the troposphere and the stratosphere.

BOX 2

Characteristics of NAM/AO, NAO, and ENSO

NAM/AO is a measure of the surface pressure gradient between the polar and subpolar regions of the Northern Hemisphere (Thompson and Wallace, 2000). When the AO index is positive, surface pressure is low in the Arctic region, which is associated with a strong North Atlantic jet (and a strengthening of the polar vortex), thus keeping cold Arctic air in the polar region. When the AO index is negative, the Arctic tends to have high pressure, weaker zonal winds, resulting in a greater movement of polar air into middle latitudes (Overland et al., 2011). The AO was strongly positive in the early 1990's compared to the previous 40 years; however, the AO has been low and variable for the past 9 years.

The NAO is characterized by changes in the North Atlantic jet stream and storm track location and intensity as well as zonal and meridional heat and moisture transport (Hurrell, 1995). The positive phase of the NAO is associated with strengthened westerlies due to an increase in the pressure gradient over the North Atlantic. Such strong westerlies allow cold air to drain off the North American continent allowing the air to flow northeast across the Atlantic into Europe. The weather for both North America and Europe is mild compared to the weather during a negative NAO. The negative phase of the NAO is characterized by a decrease in the pressure gradient across the North Atlantic, which results in a weakening of the westerlies allowing cold air to build over Canada and the eastern parts of the United States. NAO phases vary from year to year, but they can also remain in one phase for intervals lasting several years.

ENSO refers to a set of weather patterns associated with variations in sea surface temperatures (SSTs) of the tropical eastern Pacific Ocean and in the sea level pressure in the tropical western Pacific. During an El Niño phase, the tropical easterly winds decrease, warm SSTs cover the central and eastern tropical Pacific, and the heaviest rainfall moves east. El Niño is also characterized by sea-level pressure that is lower in the eastern Pacific and higher in the western Pacific. The opposite pattern occurs during a La Niña phase. In North America, El Niño winters tend to be warmer and drier than average in the northern parts of the United States, whereas northern Mexico and the southern United States typically experience wetter and cooler winters On the other hand, La Niña causes above-average precipitation across the northern United States, with precipitation in the southwestern and southeastern states typically below average. El Niño recurs at irregular intervals ranging from 2 years to a decade.

Barnes noted that the Atlantic jet stream variability is affected by the NAO and the east Atlantic pattern,[6] which are important to study in the context of Arctic linkages because they are close to the Arctic (and thus more likely to be influenced).

Rossby waves, or large meanders within the jet stream, typically travel eastward. Rossby wave breaking leads to jet stream variations, where cyclonic breaking dominates poleward of the jet and tends to drive the jet southward and anticylonic breaking dominates equatorward of the jet and tends to drive the jet northward (Benedict et al., 2004; Strong and Magnusdottir, 2008).

High-latitude wave-breaking is correlated with NAO on decadal timescales. Rossby wave breaking also serve as the mechanism for the reversal of the meridional gradient of geopotential height fields that is typical of blocks.

Atlantic blocking patterns are a critical component of NAO variability (Woollings et al., 2008). In general, a block is characterized by an atmospheric phenomenon in which a large, quasi-stationary anticyclone develops in the mid-latitudes and persists for several days or longer, blocking the ambient westerly winds and weather systems (Berrisford et al., 2007; Woollings et al., 2008). Currently, there is no consensus on exactly what type of system should be classified as a block. Blocks lead to a stagnation of weather systems and are associated with extreme weather events in the mid-latitudes (e.g. cold snaps, heat waves). They can persist for days to weeks (e.g. Black et al., 2004; Dole et al., 2011).

---

[6] The (EA) pattern is a prominent mode of low-frequency variability over the North Atlantic, appearing in all seasons except for summer. The pattern is structurally similar to the NAO. The anomaly centers of the EA pattern are displaced southeastward to the approximate nodal lines of the NAO pattern. Definition from: http://www.cpc.ncep.noaa.gov/data/teledoc/ea.shtml

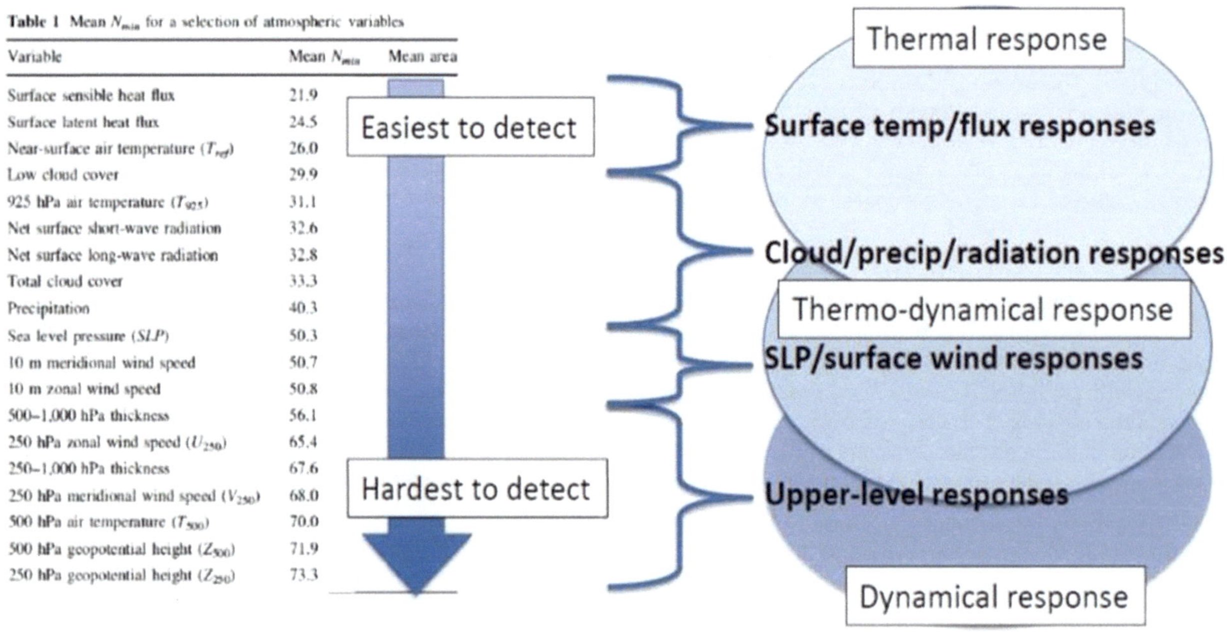

Table 1 Mean N_{min} for a selection of atmospheric variables

Variable	Mean N_{min}	Mean area
Surface sensible heat flux	21.9	
Surface latent heat flux	24.5	
Near-surface air temperature (T_{ref})	26.0	
Low cloud cover	29.9	
925 hPa air temperature (T_{925})	31.1	
Net surface short-wave radiation	32.6	
Net surface long-wave radiation	32.8	
Total cloud cover	33.3	
Precipitation	40.3	
Sea level pressure (SLP)	50.3	
10 m meridional wind speed	50.7	
10 m zonal wind speed	50.8	
500–1,000 hPa thickness	56.1	
250 hPa zonal wind speed (U_{250})	65.4	
250–1,000 hPa thickness	67.6	
250 hPa meridional wind speed (V_{250})	68.0	
500 hPa air temperature (T_{500})	70.0	
500 hPa geopotential height (Z_{500})	71.9	
250 hPa geopotential height (Z_{250})	73.3	

FIGURE 4 Atmospheric variables listed in order of ascending difficulty in detection of change. Low values imply a response that is easier to detect than high values. SOURCE: Adapted from Screen et al., 2013.

Barnes said that there are three main blocking centers in the Northern Hemisphere: the Atlantic Ocean, Europe, and eastern Asia, and most occur in winter season because of greater wave-breaking activity in winter.

Barnes stressed that jet stream shifts, Rossby wave propagation and wave breaking, and blocking events are all tightly coupled together. Mid-latitude weather variability is also dominated by interactions of the large-scale flow with synoptic-scale eddies. The latter help balance the global energy budget by "stirring" the atmosphere and moving cold air toward the equator and warm air toward the poles, which in turn, reduces the equator-to-pole temperature contrast.

Challenges in Detecting and Attributing Changes in Mid-latitude Weather

Variability of the climate system may be due to non-linear dynamical processes intrinsic to the atmosphere (internal variability; as discussed above), or to variations in natural or anthropogenic external forcing (external variability; e.g. GHG emissions, volcanoes). Barnes noted that the internal variability of the atmosphere is high, which complicates attribution of Arctic influences on the mid-latitudes. Screen et al. (2013) suggest that present-day trends in mid-latitude weather driven by Arctic change are most likely masked by internal variability and that the detection of extreme events is even more difficult (Figure 4). Some participants said that the patterns associated with a warming Arctic do not match well with any known natural variability modes (e.g., NAO, AO), which supports that idea that the recent trends are largely anthropogenic.

Barnes noted that mid-latitude jet position has a large amount of internal variability, on varying timescales (i.e., daily, yearly, decadal). Furthermore, blocking frequency is highly variable. She suggested that, given the large internal variability of the jet stream and blocking, any potential effect of Arctic warming on mid-latitude weather is unlikely to be detectable with current observations. Furthermore, Overland noted that the time series is too short (i.e., significant loss of Arctic sea ice has only been occurring for about the past 6-7 years) to robustly differentiate Arctic forcing of mid-latitude extremes from random events. Overland noted that mid-latitude attribution remains difficult and controversial because interactions between Arctic forcing and chaotic mid-latitude flow are complex. One would not expect events to happen the same way every year even with similar Arctic forcing.

Ultimately, participants noted that the issue of Arctic warming impacting regions outside the Arctic is much larger than just sea ice retreat and includes processes encapsulated under the larger umbrella of Arctic amplification. In addition, participants stressed that it is critical to assess both natural and anthropogenic mechanisms leading to sea ice loss. It is unknown to what extent rapid sea ice loss, particularly after 2005, is a part of multi-decadal oscillation or a result of Arctic warming.

Examples of Connections in the Global Climate System

Several workshop speakers presented examples of the connections and feedbacks linking the tropics, mid-latitudes, and the poles, which highlight the interconnectedness and complexity of the global climate system.

Polar Stratosphere Influence on the Mid-Latitude Troposphere

In his remarks, workshop speaker Paul Kushner, University of Toronto, discussed the coupling between the polar stratosphere and the extratropical troposphere. Stratospheric circulation variability occurs on longer time scales than tropospheric variability and can be characterized by the strength of the polar vortex, which peaks during winter in the Northern hemisphere and late spring in the Southern Hemisphere (Baldwin and Dunkerton, 2001).

Kushner said that it is well known that extratropical stratospheric variability is controlled by atmospheric composition (e.g., ozone and GHGs) as well as planetary waves (i.e. Rossby waves) propagating upward from the troposphere (Shindell et al., 1999; Sigmond et al., 2008). However, there is also "stratospheric influence", that is, variability of the stratosphere impacting tropospheric circulation on intraseasonal timescales.

Kushner discussed the pathway for this coupling, which is known as the "NAM/wave mean flow interaction pathway." Essentially, a strengthened polar vortex leads to positive AO/NAM with a 2- to 4-week lag due to wave mean flow interactions. This interaction has been well studied (i.e., Baldwin and Dunkerton, 2001; Polvani and Waugh, 2004). Kushner also discussed a second pathway for stratosphere-troposhere coupling: the "wave reflection pathway." The polar vortex can create a reflective surface for upward propagating planetary waves (e.g., Rossby waves). The reflective signal in the troposphere shows up as regional circulation anomalies such as the NAO. This connection has been less well studied, but it continues to be an active area of research (e.g., Perlwitz and Harnik, 2004; Shaw and Perlwitz, 2013).

Kushner said that it is important to examine whether the stratosphere serves as a bridge that connects Arctic sea ice loss to changes in the mid-latitude temperature, storm track, and circulation. As workshop presenter Steve Feldstein, Pennsylvania State University, noted, given the large area (i.e., large spatial scales and long timescales) influenced by Arctic sea ice anomalies, it is likely that the stratosphere plays a role.

Snow Cover Influence on the Mid-Latitude Winter Weather

Judah Cohen, Atmospheric and Environmental Research, said that Siberian snow cover has been shown to influence mid-latitude winter weather. Comparison of correlations between October Siberian snow cover and September Arctic sea ice extent and the winter AO shows that the Siberian snow cover is much more highly correlated with the winter AO than with sea ice.

Kushner said that the stratosphere may act as an intermediary in the linkage between October Eurasian snow cover and wintertime NAM/AO anomalies (e.g., Cohen and Entekhabi, 1999). This linkage likely is driven by a snow-forced planetary wave that induces stratosphere-troposphere coupling (Cohen et al., 2007; Gong et al., 2002).

Mid-Latitude Influence on the Arctic

Workshop speaker John Gyakum, McGill University, presented a case study from a synoptic meteorology perspective to illustrate how mid-latitude processes can influence Arctic weather. He discussed the apparent impacts of extratropical cyclogenesis (i.e., a low-pressure area genesis) in causing an extreme Arctic wind event along the Beaufort Sea coast in September 1999. This case study is associated with at least three extreme phenomena: (1) extreme subtropical water vapor transport; (2) extreme explosive cyclogenesis in the northeastern Pacific; and (3) extreme surface winds in the vicinity of Tuktoyaktuk, a small town on the Beaufort Sea.

This case study, Gyakum said, seems to suggest that if one component of future climate change is an eastward displacement of storm tracks into the northeastern Pacific, then more episodic events such as this one can be expected. Additionally, the decreasing coverage of Arctic sea ice may result in a longer season of Arctic cyclogenesis, with associated extreme winds and coastal erosion.

Tropical Influence on the Arctic

In his remarks, Feldstein presented evidence of the tropics playing a role in Arctic amplification. He noted that there is some disagreement whether Arctic amplification is driven by sea-ice albedo feedback or by poleward transport of heat from the lower latitudes. There has been much research on the former, but relatively little on the latter. The Tropically-Excited Arctic warMing (TEAM) is a proposed mechanism for tropical convection, Arctic amplification, and a reduction of Arctic sea ice. The trend toward stronger and more localized convection over the tropical Indo-Pacific Warm Pool results in the propagation of Rossby waves toward the poles. These Rossby wave trains warm the Arctic by transporting heat and moisture poleward, inducing sinking motion/adiabatic warming[7] over the Arctic and increasing the downward flux of infra-red radiation (Lee et al., 2011a; Lee et al., 2011b; Yoo et al., 2011; Yoo et al., 2012a; Yoo et al., 2012b). Dargan Frierson, University of Washington, also discussed the transport of heat to the Arctic but noted that some modeling studies suggest that heat transport to the Arctic is reduced as the Arctic becomes warmer (Appendix A).

Arctic Influences on the Climate System

Arctic Influence on the Tropics

In his remarks, Frierson noted that model simulations show that Arctic regions affect climate elsewhere. For example, Arctic changes can affect tropical rainfall. Work by Chiang and

[7] A process whereby the temperature changes because of an expansion or compression.

Bitz (2005) demonstrates a strong sensitivity of tropical rain bands to Arctic sea ice increases. Rain bands shift southward as a result of the subsequent cooling in the northern high- and mid-latitudes because of the increased sea ice. Climate changes in the extratropics also influence the storm tracks over the Southern Ocean.

Furthermore, Frierson noted, there is evidence of "interhemispheric teleconnections". A recent study shows a poleward shift of the Southern Hemisphere jet stream in response to Northern Hemisphere extratropical cooling alone (Figure 5; Ceppi et al., 2013).

Arctic Influence on the Mid-Latitudes: A Synoptic Meteorology Perspective

Gyakum presented a second case study, which illustrates how Arctic processes may have led to the formation of a mid-latitude storm. He discussed the apparent impacts of Arctic air mass formation on explosive extratropical cyclogenesis in January-February 1979. Northwestern Canada is a primary location for the formation of air masses that impact North America and is also a region of significant cold-season warming during the past several decades. Deep-tropospheric cold air regions facilitate triggers for cyclogenesis in the form of synoptic-scale troughs. This case study highlights Arctic air mass formation mechanisms and possible long-term changes in such mechanisms.

Arctic Influence on the Hydrological Cycle in the Mid-Latitudes and Tropics

Ian Eisenman, University of California, San Diego, said that change in surface albedo (due to Arctic sea ice loss) has been suggested to affect the hydrological cycle in the mid-latitudes and tropics. Through modeling studies he found that ice albedo impacts the global hydrological cycle in cold climates, but not in warm climates, with the transition occurring at a point relatively near to the present-day climate. He said these results suggest that ice albedo effects are important for understanding the hydrological cycle in climates colder than today, including many paleoclimate environments.

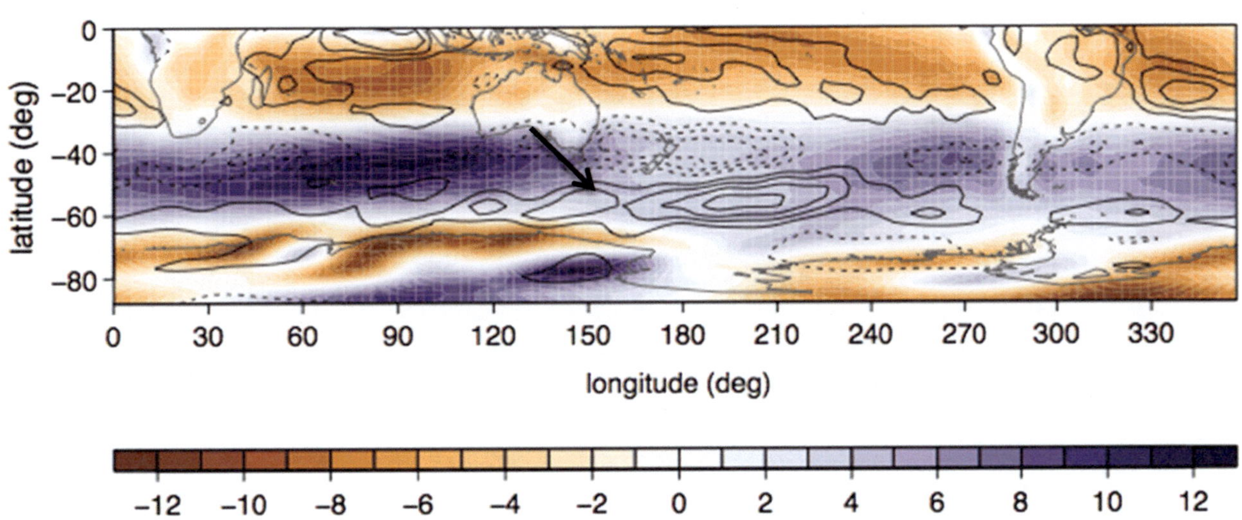

FIGURE 5 Surface zonal wind change (contours) and control (shading) from cooling at 50 N. SOURCE: Ceppi et al., 2013

Arctic Influence on Ocean Circulation

Arctic change indirectly influences the mid-latitudes through changes in ocean stratification and circulation. Many workshop participants asserted that understanding how longer-term ice loss will impact ocean circulation and eventually mid-latitude weather is critical. In her remarks, workshop speaker Marika Holland, NCAR, presented modeling studies that indicate that freshwater inflow to the Arctic Ocean due to sea ice melt plays an important role in controlling the deep North Atlantic ocean convection and ocean meridional circulation, which in turn affects global climate (Kug et al., 2010). Modeling studies project that significant loss of Arctic sea ice will reduce solid (ice) transport to the North Atlantic but will increase liquid transport to the North Atlantic (Jahn and Holland, 2013). This increase in freshwater liquid transport to the North Atlantic could have implications on North Atlantic circulation (decrease in the Atlantic meridional overturning circulation [AMOC]) resulting in reduced ocean heat transport and reduced warming locally. Regarding the impact of longer-term ice-loss changes on mid-latitude weather, these ocean changes must be considered, she said.

Holland also discussed work by Laura Landrum, NCAR, and colleagues that showed the oceanic response to Arctic sea ice loss. They ran a summer ice free Arctic simulation with summer ice loss forced through albedo modifications and no greenhouse-gas changes. The results show a decrease in AMOC in the first 40 years of the ice free simulation, which is (largely) attributable to ice loss.

Oceanic Feedbacks Influence on Arctic Amplification

Clara Deser, NCAR, discussed the role of oceanic feedbacks in the atmospheric response to GHG-induced Arctic sea ice loss. Her model results show that the primary effect of having an interactive ocean is to deepen the atmospheric temperature response to the mid-troposphere (about 400 hPa) compared to the atmosphere-only run in which the thermal response is confined to the planetary boundary layer (below 850 hPa). This result highlights the contribution of high-latitude ocean feedbacks to Arctic amplification. According to Deser, the near-surface atmospheric circulation response did not appear to be sensitive to ocean coupling.

Conclusion

From this big picture context, where Arctic sea ice, atmosphere, and ocean processes are tightly coupled and are all relevant controls on mid-latitude weather, many participants stressed the importance of understanding the various processes on a global scale and how they interact with one another. This in turn is necessary for understanding not only the impact of Arctic sea ice loss on mid-latitude weather, but also more generally the interactions between Arctic and mid-/low-latitude processes (i.e., the pushing and pulling forces on the mid-latitudes). Furthermore, participants noted that it is evident from the workshop that this issue is much larger than Arctic sea ice loss (the title of the workshop). Rather, it would be more appropriate to place focus on the processes encapsulated under Arctic amplification.

Some of the workshop speakers urged participants to think systematically about how Arctic changes might directly influence the mid-latitudes and about how changes in mid-latitudes might be affected by the Arctic through feedbacks. They discussed components of the climate system as we currently understand it while acknowledging complicating factors such as interacting systems and feedbacks that make attribution difficult. For example, some participants pointed out that any change that projects onto modes of variability (e.g., AO/NAM) may simultaneously affect the Arctic and mid-latitudes.

OBSERVATIONAL EVIDENCE OF TRENDS

One goal of the workshop was to discuss observational evidence of possible Arctic linkages. This section highlights observational evidence of some mechanisms that were raised at the workshop as possibly linking Arctic amplification and mid-latitude weather:

- Arctic warming faster than the Northern Hemisphere
- Decrease in the temperature gradient between the Arctic and the mid-latitudes
- Slowing of upper-level zonal winds
- Upper-level flow becoming more meridional
- Increase in the amplitude of large-scale waves
- Increase in blocking events
- Large-scale waves progress more slowly eastward
- Increase in extreme events
- Weakening of the polar vortex

Arctic Warming Faster Than the Northern Hemisphere

As discussed in the previous section, the Arctic is warming faster than the tropics and mid-latitudes. This has resulted in a greater flux of energy from the newly exposed open ocean (relative to ice-covered waters). The strong Arctic warming is not confined to a shallow surface layer, but extends through a sufficiently deep layer of the atmosphere to influence geopotential thicknesses. In his remarks, Overland presented evidence of anomalous temperature patterns in the Northern Hemisphere (Figure 6). He acknowledged that we cannot prove that these changes are due to changes in the Arctic sea ice loss, but something unusual is occurring.

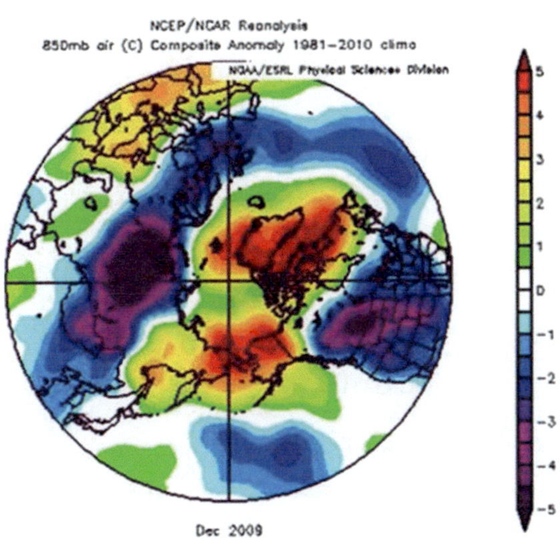

FIGURE 6 Anomolous temperatures in the Northern Hemisphere. Figure is from December 2009, which had a record negative AO. Note the below-average temperatures throughout the United States and the above-average temperatures around the Arctic. SOURCE: NCEP/NCAR.

In her remarks, Jennifer Francis, Rutgers University, said that current Arctic amplification is strongest in autumn and winter and is a relatively recent phenomenon, which begins to appear in the mid-1990s. She said that this short time series makes it difficult to find a statistical trend in the observations to support the existence of Arctic linkages. Others in the room disagreed that Arctic amplification recently emerged.

Decrease in the Poleward Temperature Gradient

It has been suggested that the increase in the flux of energy from the newly exposed open ocean has led to the weakening of the poleward temperature gradient (i.e., a decline in the temperature contrast between the Arctic and mid-latitudes). Francis presented observational evidence of this in Figure 7.

Overland discussed how upper-level atmospheric circulation in north-central Asia in winter responded to the recent large-scale reduction in the north-south temperature gradients, reducing jet-stream zonal velocities. Overland believes that the reduction in jet-stream zonal velocities led to the penetration of storms into northern Asia from the west, increasing the strength and persistence of the Siberian High, and thus increasing the intra-seasonal probability of multiple cold air events over eastern Asia.

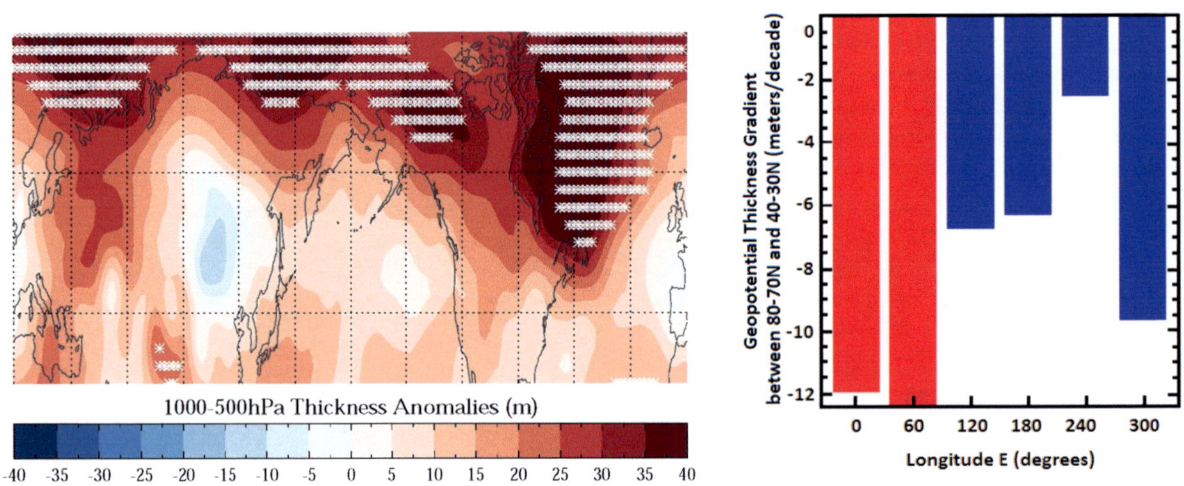

FIGURE 7 A negative trend in poleward geopotential thickness gradient (1000-500 hPa) from 1979 to 2012. Left: Change in geopotential thickness in 1000-500 hPa during 1979–2012 in autumn (October/November/December [OND]). Statistically significant changes in geopotential thickness (the white X's) are apparent over much of the Arctic regions. Right: Trends in poleward geopotential thickness gradient (1000-500 hPa) between 80-70 N and 40-30 N in units of meters per decade corresponding to longitude. The time period is from 1979 to 2012. A value of -10 m/decade is about -2 percent per decade. The red bars are significant at the 90 percent confidence level. Places where the geopotential thickness anomalies are greatest (i.e., where the temperature gradient has weakened the most; left figure[8]) correspond with a negative trend in poleward geopotential thickness gradient (red bars in the right figure). This signal is strongest in autumn and more variable in summer. Data are from the NCEP/NCAR Reanalysis. SOURCE: Jennifer Francis, Rutgers University.

[8] The top figure shows geopotential thicknesses, not temperatures, although a greater temperature will result in a larger 1000-500 hPa distance.

Upper-Level Zonal Winds Decreasing Where Temperature Gradient Weakens

Some participants said that reduced poleward temperature gradients should result in weakened westerlies. Overland believes that the similar penetrations of cold air into the eastern United States in Decembers 2009, 2010, and 2012 relate to a shift in the long-wave upper-level atmospheric wind pattern (Figure 8). This shift coincides with warmer temperatures and greater geopotential thickness over northeastern Canada, major sea ice loss during October in Baffin Bay, a positive Greenland Blocking Index (greater 500 hPa geopotential heights), and record negative values of the AO index. Such a combination of events amplified and shifted the climatological atmospheric wind pattern westward and allowed deeper southward penetration of cold air into the United States. Northward airflow over Davis Strait acted as a positive feedback to maintain the higher air temperature anomalies. He acknowledged that this is not statistically significant given that this pattern occurred 3 years out of 4, but suggests the wind shift may be the start of a trend.

Francis presented evidence of zonal winds at 500 hPa (roughly 5,500m above sea level) weakening in autumn (OND) and winter (January/February/March [JFM]; Figure 9). Figure 10 shows evidence of the upper-level zonal winds decreasing in areas where the temperature gradient has also weakened. This is more apparent in autumn, but also occurs in summer. John Walsh also presented evidence on zonal wind speeds at 500hPa (Figure 11). They weakened from 1979 to present in autumn (OND), but if observations are extended back to 1948, then zonal wind speed increases from the 1950s to the late 1970s. He noted that there is no known corresponding trend of increasing sea ice cover during the summer/autumn period in these decades. Francis noted that Walsh considered 30-80°N, whereas she took a more regional focus (40-60°N).

Upper-Level Flow Becoming More Meridional

Another proposed mechanism is upper level winds becoming more meridional flow due to the decrease in wind speeds. Francis presented evidence of this occurring in autumn (OND; Figure 12).

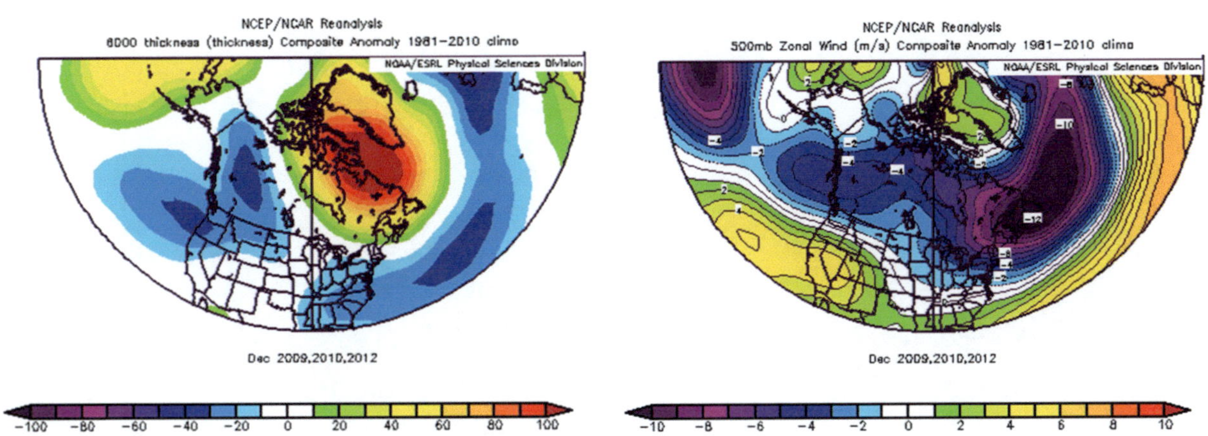

FIGURE 8 Left: Composite 1000-500 hPa geopotential thickness anomaly field for Decembers 2009, 2010, and 2012. Right: 500 hPa zonal wind anomaly for Decembers 2009, 2010, 2012. Weakened westerlies, especially in the northwest Atlantic, help support the idea that the extraordinary warming over Baffin Bay led to a more meandering jet stream. Data downloaded from the NCEP atlas. Data from NCEP/NCAR reanalysis. SOURCE: James Overland, NOAA.

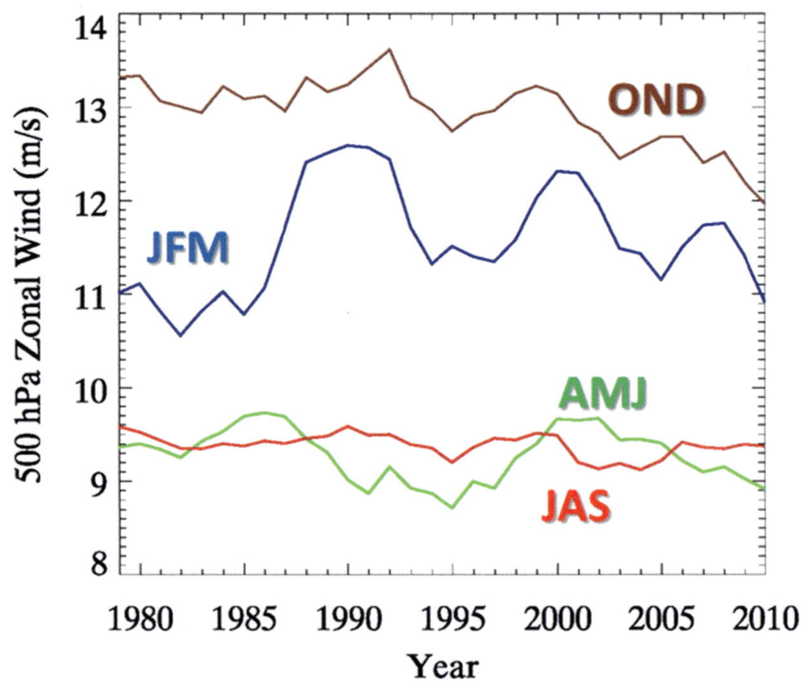

FIGURE 9 Seasonal zonal wind speed at 500 hPa from 1980 to 2010. Green line: spring (April/May/June [AMJ]); red line: summer (July/August/September [JAS]); blue line: winter (JFM); brown line: autumn (OND). SOURCE: Jennifer Francis, Rutgers University

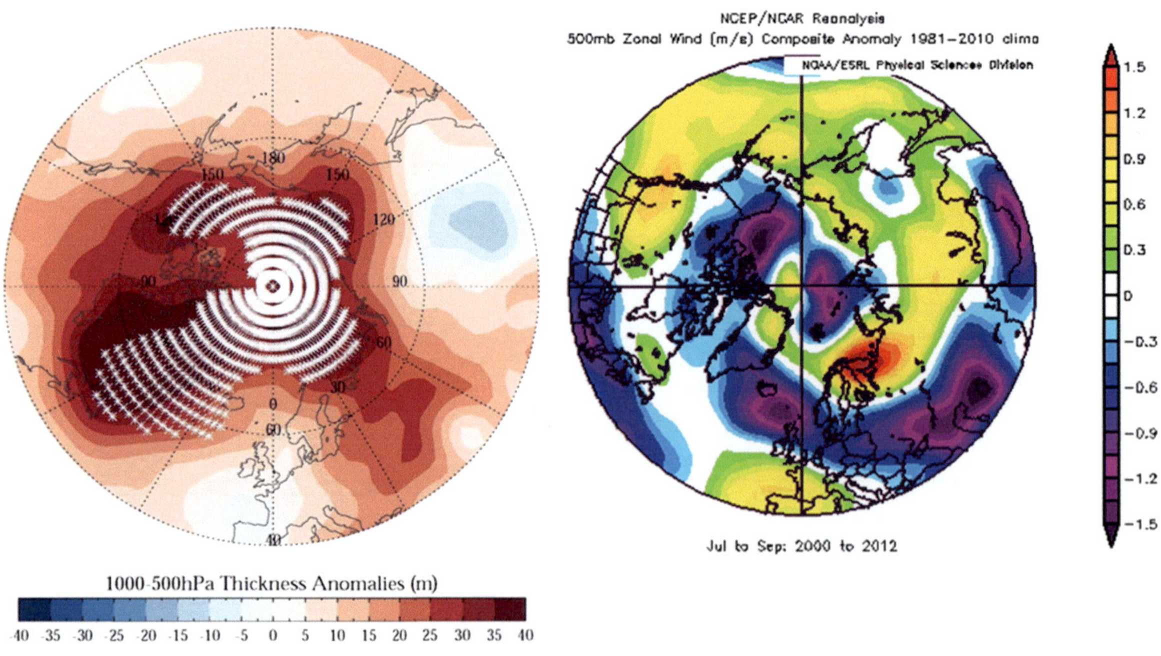

FIGURE 10 Left: 1000-500 hPa geopotential thickness anomalies in autumn (OND) from 2000 to 2012. Right: Anomalies in zonal winds at 500 hPa from 2000 to 2012. Francis noted that areas where there is a decrease in zonal winds (500 hPa) correspond to areas in the Northern Hemisphere where there are weakening temperature gradients. SOURCE: Jennifer Francis, data from NCEP/NCAR.

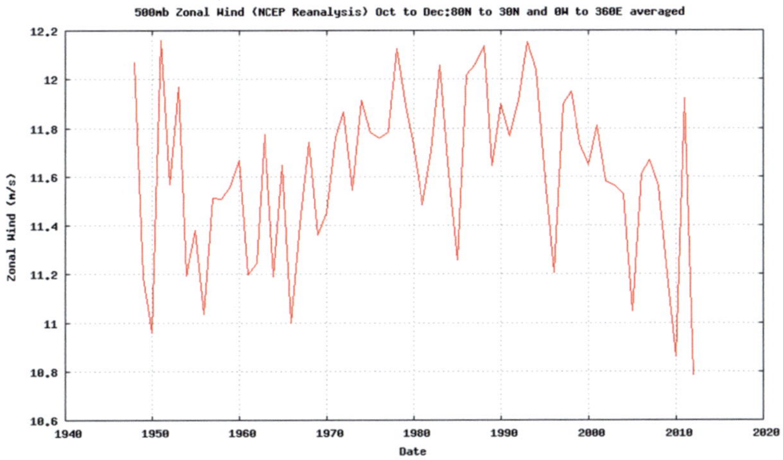

FIGURE 11 Autumn (OND) mean zonal wind at 500 hPa (30-80ºN, 0-360ºW), 1948-2012. Data from NCEP. SOURCE: John Walsh, University of Alaska, Fairbanks.

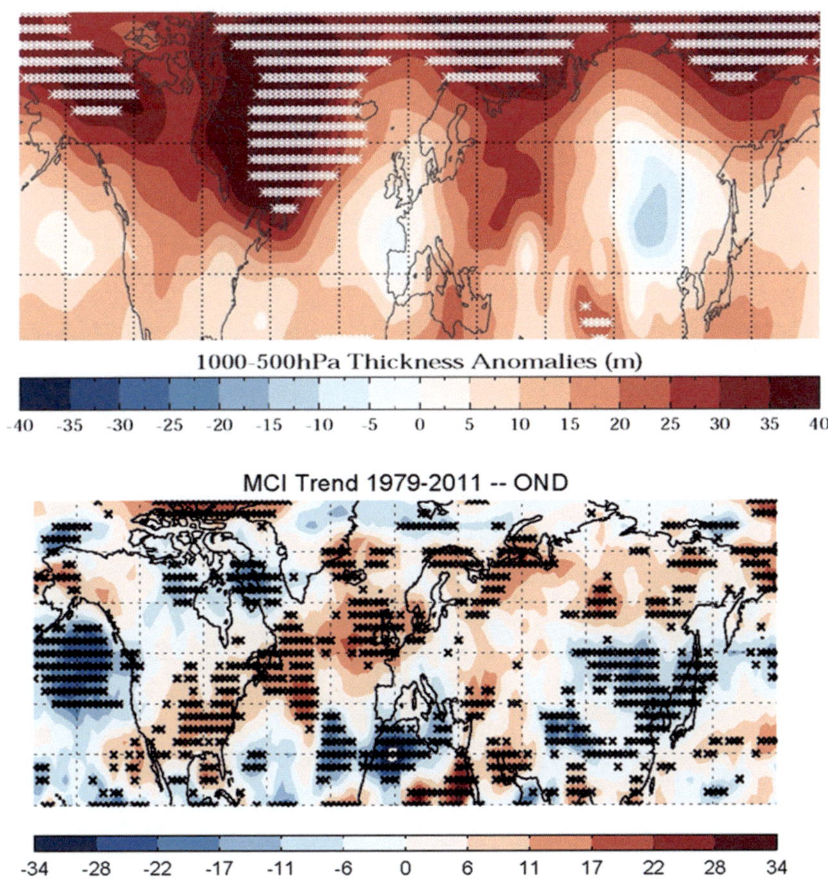

FIGURE 12 Trends in meridional component of the 500 hPa wind in autumn (OND; 1979-2011). Top: 1000-500 hPA change in geopotential thickness in OND. Bottom: Trend in waviness in OND from 1979 to 2011 in the Northern Hemisphere. Blue is less wavy and red is more wavy. Francis noted that the areas with a larger change in geopotential thickness (top) correspond well with the pattern of areas with a more meridional flow (bottom). She also noted that there is a less cohesive pattern in summer (JAS), but there is still a correspondence between a weakening temperature gradient and a more meandering upper-level flow.

Some participants said that if there is a statistically significant slowing of waves and weakening of westerlies, then we still should acknowledge that this trend is not necessarily caused by one single mechanism. In addition, it is critical that we can point to other competing mechanisms, not just random "noise" (natural variability).

Increase in Rossby Wave Amplitude

It has also been suggested that the amplitude of Rossby waves is increasing (i.e., an increase in the north-south extent of the waves), which coupled with the decrease in zonal winds, results in a jet stream that is more meandering. Francis said there is a positive trend in the size, or amplitude of the waves, in autumn (OND; Figure 13). She also presented evidence that there is a positive trend in the frequency of ridges according to longitude.

Both Screen and Barnes presented evidence from Screen and Simmonds (2013) and Barnes (2013) that suggest that amplitude trends are metric dependent. Francis also noted this variable is difficult to measure objectively. Screen and Simmonds (2013) measured wave amplitude differently from Francis and Vavrus (2012) and found significant increases in meridional wave amplitude over Europe during spring (AMJ), but not in any other months or seasons. They also found significant decreases in zonal amplitude (a measure of the intensity of atmospheric ridges and troughs at 45 N) in the entire Northern Hemisphere and also individually over Europe and Asia (Figure 14). Screen also noted that they found statistically insignificant positive trends in all seasons in the North America and North Atlantic regions, which is in contrast to the comparatively larger (and significant) increases in summer (JAS) and autumn (OND) found by Francis and Vavrus (2012).

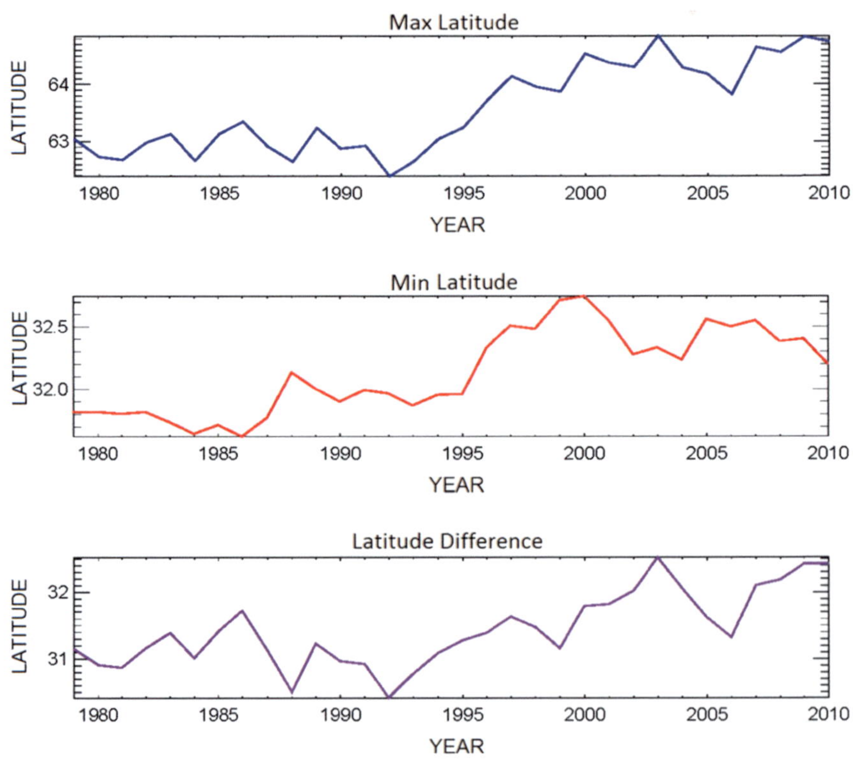

FIGURE 13 Large-scale wave trends in autumn (OND) in the Northern Hemisphere. Top: Latitude of peak of ridges. Middle: Latitude of base of troughs. Bottom: The difference between the peak of ridges and the base of troughs (i.e., a measure of wave amplitude). Amplitude of large-scale waves appears to be increasing since the early 1990s. SOURCE: Jennifer Francis, Rutgers University.

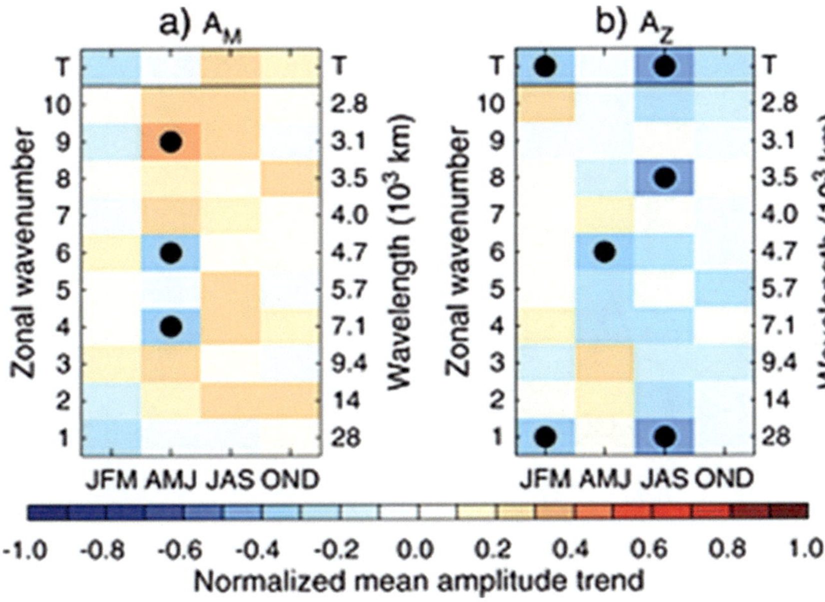

FIGURE 14 Linear trends from 1979 to 2011 at 500 hPa geopotential height in seasonally averaged meridional amplitude. Statistically significant trends are identified by the black dots. The trends in meridional amplitude are positive in summer (JAS) and autumn (OND) and negative in winter (JFM) and spring (AMJ), but none of these is significant. Only three of the trends for individual wavelengths are statistically significant. Left: Meridional amplitude. Right: Zonal amplitude. SOURCE: Screen and Simmonds, 2013.

Increase in Blocks

It has been proposed that blocking trends will be more likely to increase due to the increase in Rossby wave amplitude. Barnes noted that the multiple studies on observed blocking trends have had mixed results, and that it is likely that results are dependent on the methods used by the authors. Francis presented evidence of an increase in blocking events in autumn (OND) in the Northern Hemisphere from 1980 to 2010 (Figure 15). Barnes noted that some studies show that high-latitude blocking over the North Atlantic has decreased in the past 40 years (Barnes, 2013; Croci-Maspoli et al., 2007; Davini et al., 2012; Figure 16) whereas other studies show that low-latitude blocking has increased (Croci-Maspoli et al., 2007; Davini et al., 2012). Barnes (2013) did not find an increase in blocking trends, however, she did point out that detection can be dependent on the algorithm used, so additional studies are needed.

Barnes suggested that reported trends (from Francis and Vavrus, 2012) in wave amplitude do not appear robust because no significant trends are found when (1) daily wave extents are analyzed instead of seasonal maxima and minima and (2) a larger range of isopleths are analyzed. A poleward shift of the isopleths with Arctic amplification may appear as a change in wave extent when a narrower range of isopleths is used instead (Barnes, 2013).

Large-Scale Waves Progress More Slowly Eastward

It has been proposed that both the weakening of west-to-east upper-level winds and the more meandering trajectory of the jet stream have resulted in large-scale waves in the jet stream to progress more slowly eastward. Francis presented evidence of this slowing (Figure 17), but she noted more work is needed to assess changes in the speed of large-scale wave progression.

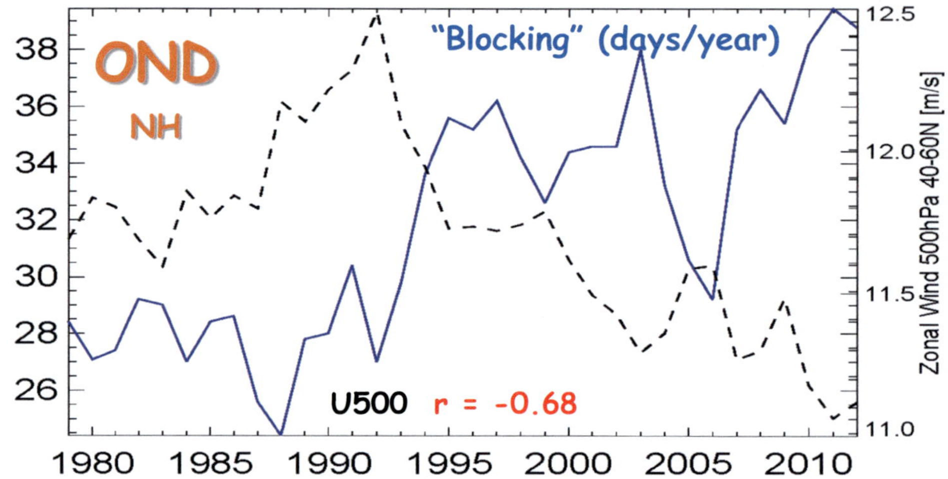

FIGURE 15 Blocking events have increased while zonal winds have decreased. Blocking events (days/year; blue line) plotted with zonal wind speed at 500 hPa (dashed line) in autumn (OND) in the Northern Hemisphere from 1980 to 2010. SOURCE: Jennifer Francis, Rutgers University.

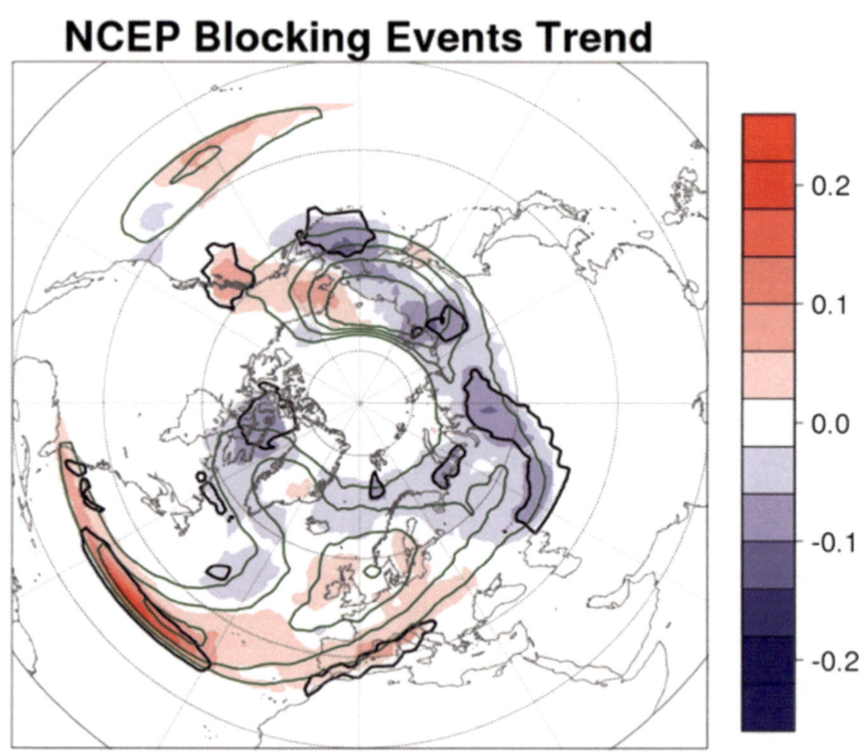

FIGURE 16 NCEP blocking trends from 1951 to 2010. Blue areas experienced a decrease in blocking events; red areas experienced an increase in blocking events. SOURCE: Davini et al., 2012.

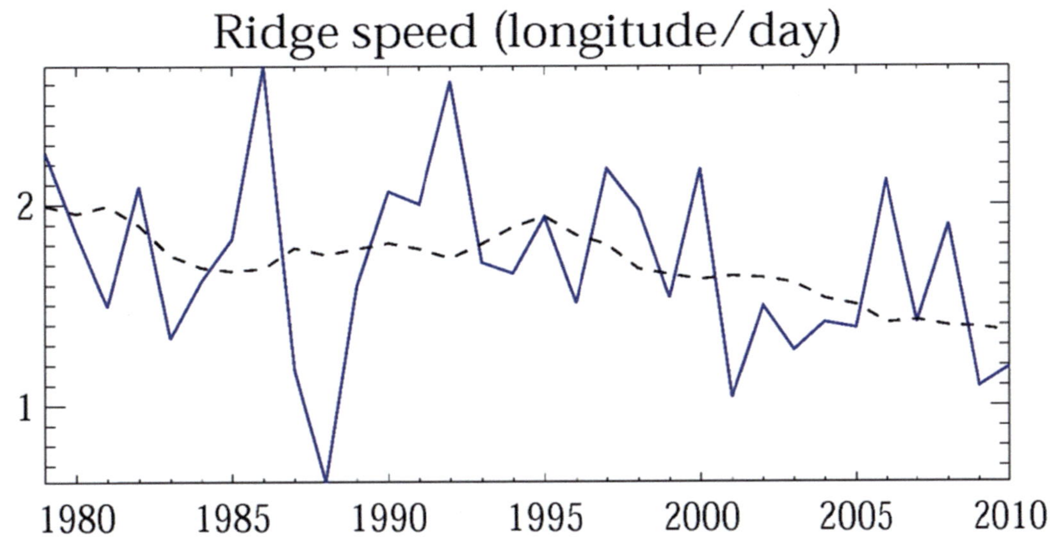

FIGURE 17 The speed of ridges (longitude/day) from 1980 to 2010, which appears to be decreasing. SOURCE: Jennifer Francis.

In agreement with Francis, Barnes (2013) found that Rossby wave speeds at 500 hPa decreased in OND. However, she found no decrease when 250 hPa meridional wind is used from 1980 to 2011 (Figure 18). Francis noted that this level is near the tropopause, often above the jet stream, and can be affected by dynamics of the stratosphere.

Barnes stressed that relationships between Rossby wave propagation and zonal wind speeds are complex. She believes that the hypotheses put forth (e.g., of Francis and others) could be correct, but there are not enough data to accurately detect trends.

More Persistent Weather Patterns, Extremes More Likely

Some workshop participants noted that weather conditions associated with the slower large-scale waves become more persistent because of an increase in blocks, increasing the probability of the types of extreme weather associated with long-lived weather conditions.

Francis noted that this variable is difficult to measure. She presented work from Vavrus on upper-air circulation anomalies during extreme weather in Chicago (Figure 19). She also discussed a regression example from Tang et al (2013), which shows that winter sea ice anomalies have a strong association with extreme cold events over the United States.

Weakening of the Polar Vortex

Workshop participants briefly discussed the hypothesis that Arctic sea ice loss leads to changes in regional heat and other energy fluxes, which result in a weak (or unstable) polar vortex. "Normal" strong polar vortex winds, which circle the Arctic from west to east, isolate cold polar air from the mid-latitudes in winter (Figure 20). Thus, it is typically mild across the eastern United States, Europe, and East Asia during winters when the polar vortex is strong. A weak or unstable polar vortex has a more north and south meandering pattern (rather than west to east). This allows cold air from the Arctic to spill into the mid-latitudes and warm air from the subtropics to move into the Arctic. It is typically colder across the eastern United States, Europe, and East Asia during winters when the polar vortex is weak.

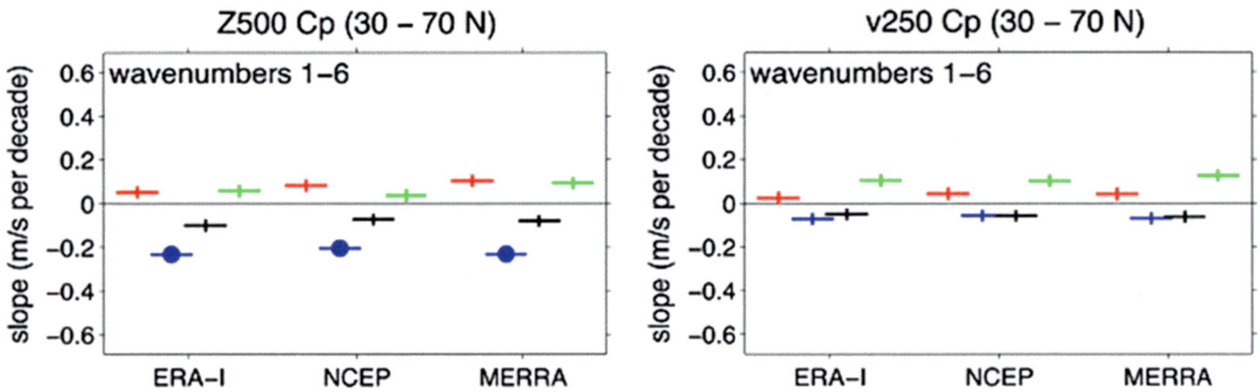

FIGURE 18 Seasonal trends of (left) Z500 phase speeds and (right) v250 phase speeds. All trends are for the North America and North Atlantic regions, and averages are taken between 30 N and 70 N. Closed circles denote trends that are statistically different from 0 at 95% confidence. Blue: autumn (OND); red: summer (JAS); black: winter (JFM); green: spring (AMJ). SOURCE: Barnes, 2013

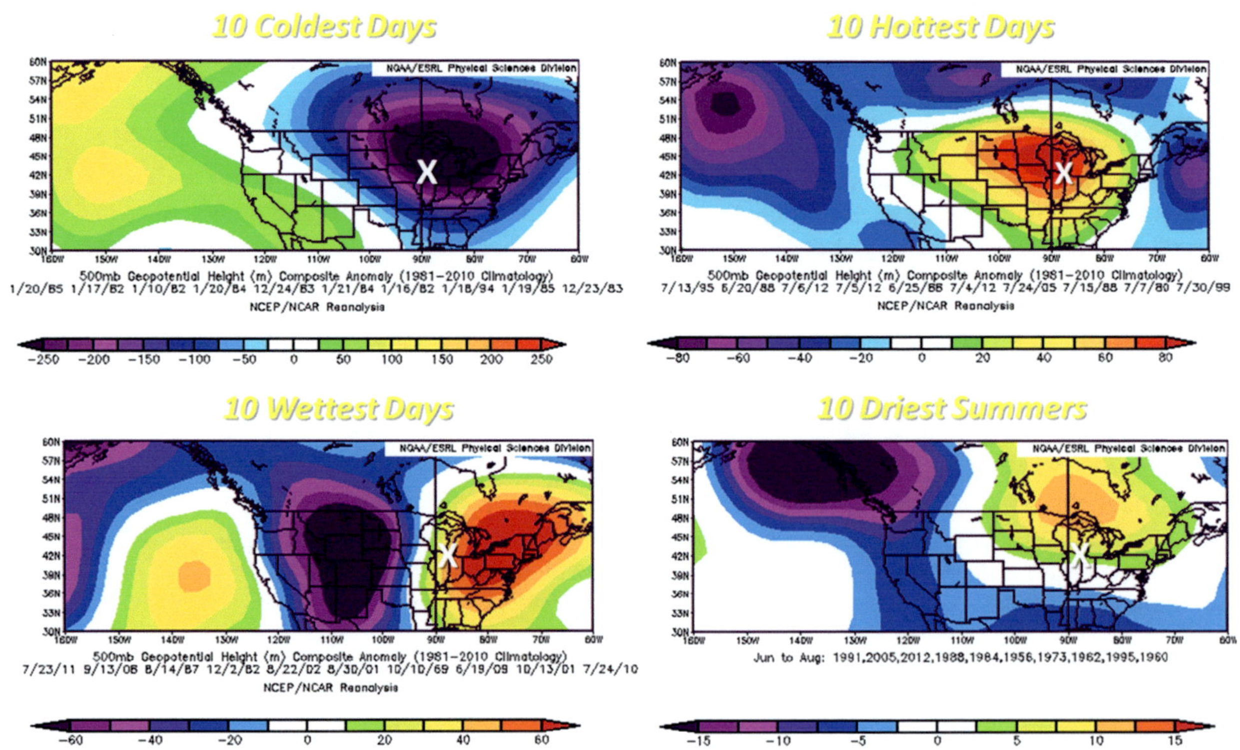

FIGURE 19 Upper-air circulation anomalies in Chicago during 10 coldest days (upper left), 10 hottest days (upper right), 10 wettest days (lower left), and 10 driest summers (lower right). Note the location of the trough (purple) during the coldest days and the location of the ridge (red) during the hottest days. SOURCE: Jennifer Francis.

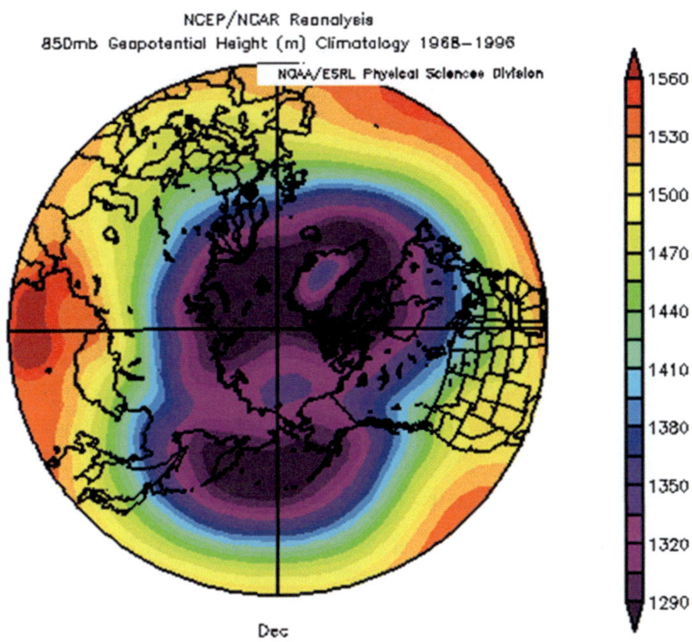

FIGURE 20 The climatological 850 hPa geopotential height field for December from 1968 to 1996. The map is indicative of normal early winter atmospheric conditions. Note the low geopotential heights of constant pressure surfaces over the Arctic (purples). SOURCE: http://www.arctic.noaa.gov

Overland and Wang (2010) find evidence that loss of sea ice and the consequent changes in regional heat and other energy fluxes can weaken the polar vortex. Continued loss of snow and sea ice adds additional heat to the atmosphere, increasing the chance of a breakdown of the polar vortex through the thermal wind mechanism. In the Pacific Arctic, reanalysis shows 2005-2011 autumn temperature anomalies reaching the middle troposphere, which supports the fact that sea ice loss impacts the larger atmospheric climate (Figure 21).

Conclusion

Some of the observational evidence presented supports hypotheses for linking Arctic amplification and mid-latitude weather patterns. Other evidence does not. Observations can help decipher whether there is a trend, but some participants noted that there may be several mechanisms behind one trend. Furthermore, finding observational trends in some of the mechanisms (wave amplitude, blocks, and wind speed) is largely dependent on how they are measured and defined.

Some participants noted that the short time series of observations of the recent significant loss of sea ice is also a limiting factor (less than 10 years) in finding trends to support the existence of these linkages and to robustly differentiate Arctic forcing of mid-latitude extremes from random events. Overland noted that the external forcing of Arctic sea ice loss and snow loss occurs on local scales and on short timescales, so, in his opinion, taking zonal averages and seasonal averages will smooth out some of the effects that we would be looking for in trying to identify trends.

Francis acknowledged more work is needed to assess the following:

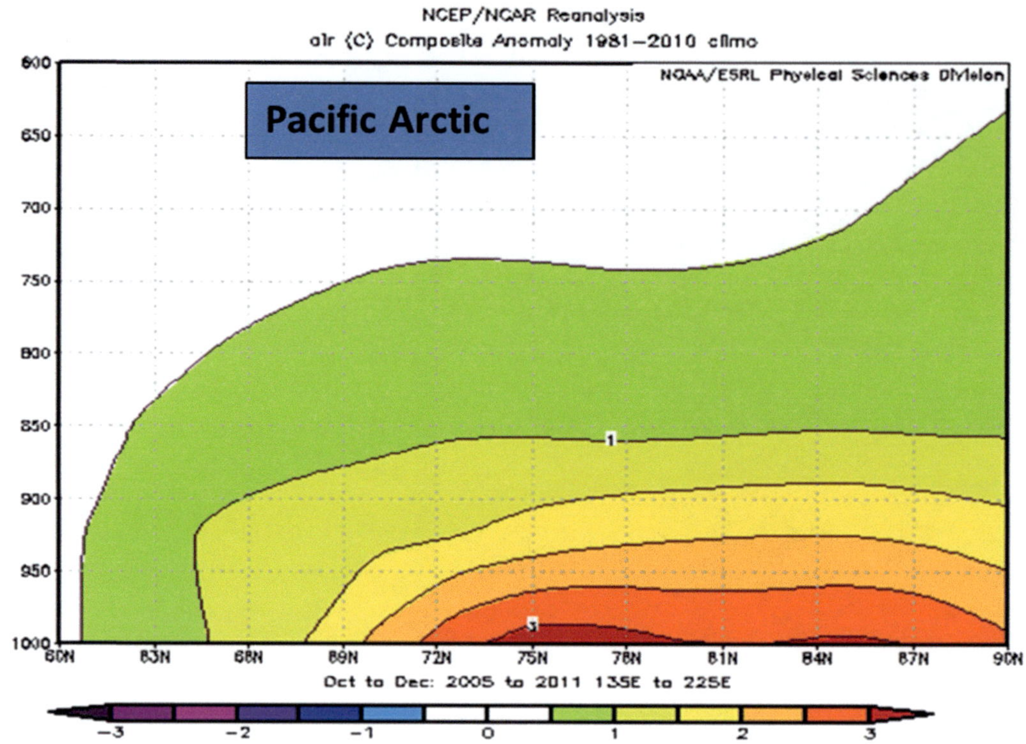

FIGURE 21 2005-2011 autumn temperature anomalies reach the middle troposphere. SOURCE: NCEP/NOAA.

- Changing propagation of large-scale waves (speed and cause)
- Changing persistence of weather patterns
- Changing frequency of extremes
- Interactions among Arctic amplification and other large-scale influences (ENSO, Pacific Decadal Oscillation [PDO], etc.)

THEORETICAL AND MODELING STUDIES

Another goal of the workshop was to explore theoretical and modeling work to test the mechanisms of proposed linkages. Many participants noted that the model spread[9] can offer clues to mechanisms and that models can assess a large number of possible pathways to influence weather because they can simulate complex large-scale dynamics reasonably well. Some participants said that, given the natural variability of the atmosphere and the numerous external factors that influence it, modeling studies are useful for identifying the contributions of each of those external factors. The following variables were discussed in the context of models that were forced with Arctic sea ice loss or Arctic warming:

- temperature gradients,
- upper-level zonal winds,
- large-scale wave amplitudes,
- blocking, and
- weather patterns and storm tracks.

[9] The degree to which models agree with one another. It is often quantified from the across-model standard deviation.

Temperature Gradients

Gudrun Magnusdottir, University of California, Irvine, focused her remarks on a recent modeling study (Peings and Magnusdottir, 2014). The researchers used NCAR's CAM5 (an atmospheric global climate model) to understand the effect of recent extensive sea-ice loss years (2007-2012; denoted as 2010C) on the atmospheric circulation in winter and then compared those results to response to projected sea-ice change at the end of the century (2080-99; denoted as 2090C). The control experiment (CTL) is a 50-year simulation with sea ice concentration representative of the 1979-2000 period. One result from the study shows that the high-latitude surface warming is greater in 2090C than in 2010C and is significant up to the 500-hPa level. This leads to increased geopotential thickness of the lower troposphere (Figure 22), which was also detected in observations by Francis and Vavrus (2012). The thermal expansion of the Arctic troposphere reduces the meridional temperature and geopotential thickness gradients from the pole to mid-latitudes from 1000 to 500 hPa.

However, Barnes said that model simulations of change in zonal mean temperature in 25 CMIP5[10] models under RCP8.5[11] find that the upper-level tropospheric (250 hPa) temperature gradient is projected to increase in all seasons. Near-surface temperature gradient is projected to decrease in the cool months with Arctic amplification. The largest uncertainty (in degrees) is found during the months of December, January, and February on the polar surface.

Upper-Level Zonal Winds

Some modeling studies discussed show a decrease in upper-level zonal winds. Francis presented a CCSM4 run in a 4xCO_2 scenario. The run shows that zonal winds are projected to decrease in all four seasons in the future (Figure 23).

Similarly, Magnusdottir noted that Peings and Magnusdottir (2014) found a weakening of the zonal westerly flow in the 2010C run that is due to the decrease in the meridional gradient of atmospheric geopotential thickness between the pole and mid-latitudes.

Large-Scale Wave Amplitudes

Screen and Magnusdottir noted that Peings and Magnusdottir (2014) found weak but significant increases in winter (DJF) mean wave amplitude under 2090 sea ice forcing. However, there was no significant wave amplitude change under 2010 sea ice forcing (Figure 24).

---

[10] In 2008, the World Climate Research Programme (WCRP) Working Group on Coupled Modelling (WGCM), with input from the International Geosphere-Biosphere Programme (IGBP) AIMES project, agreed to promote a new set of coordinated climate model experiments. These experiments comprise the fifth phase of the Coupled Model Intercomparison Project (CMIP5). CMIP5 will provide a multi-model context for (1) assessing the mechanisms responsible for model differences in poorly understood feedbacks associated with the carbon cycle and with clouds, (2) examining climate "predictability" and exploring the ability of models to predict climate on decadal time scales, and, more generally, (3) determining why similarly forced models produce a range of responses. (http://cmip-pcmdi.llnl.gov/cmip5/)

[11] Representative Concentration Pathways (RCP) are GHG concentration trajectories used for climate modeling and research. The numbers associated with RCPs (RCP8.5, RCP6, RCP4.5, and RCP2.6) refer to radiative forcings (global energy imbalances), measured in watts per square meter, by the year 2100. The four RCPs include one mitigation scenario leading to a very low forcing level (RCP2.6), two medium stabilization scenarios (RCP4.5 and RCP6) and one very high baseline emission scenarios (RCP8.5; van Vuuren et al., 2011).

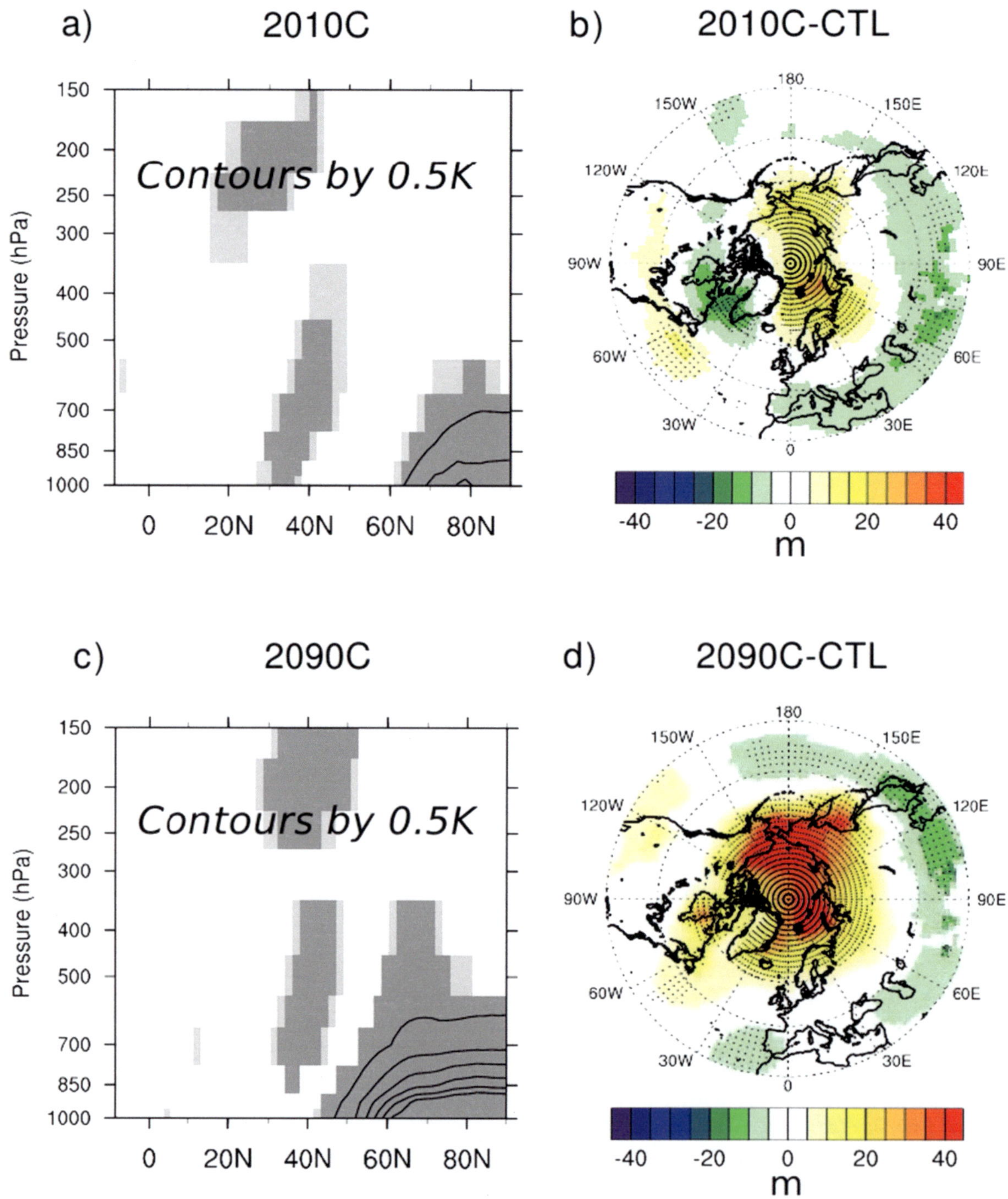

FIGURE 22 Increased geopotential thickness of the lower troposphere due to sea ice loss. (a) 2010C zonal mean DJF temperature response (K). (b) 2010C-CTL response of the DJF atmospheric geopotential thickness between 1000 and 500 hPa; (c) 2090C zonal mean DJF temperature response. Contour interval 0.5 K; light (dark) shading indicates the 90 percent (95 percent) significance level. (d) 2090C-CTL response of the DJF atmospheric geopotential thickness between 1000 and 500 hPa. SOURCE: Peings and Magnusdottir, 2014.

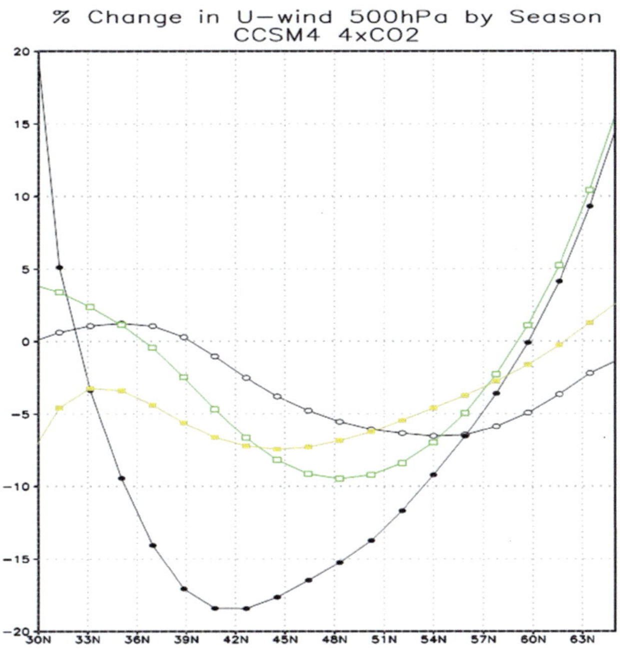

FIGURE 23 Percentage change in 500hPa zonal winds in 4xCO_2 run of CCSM4. Black (summer), green (spring), yellow (autumn), blue (winter). SOURCE: Jennifer Francis.

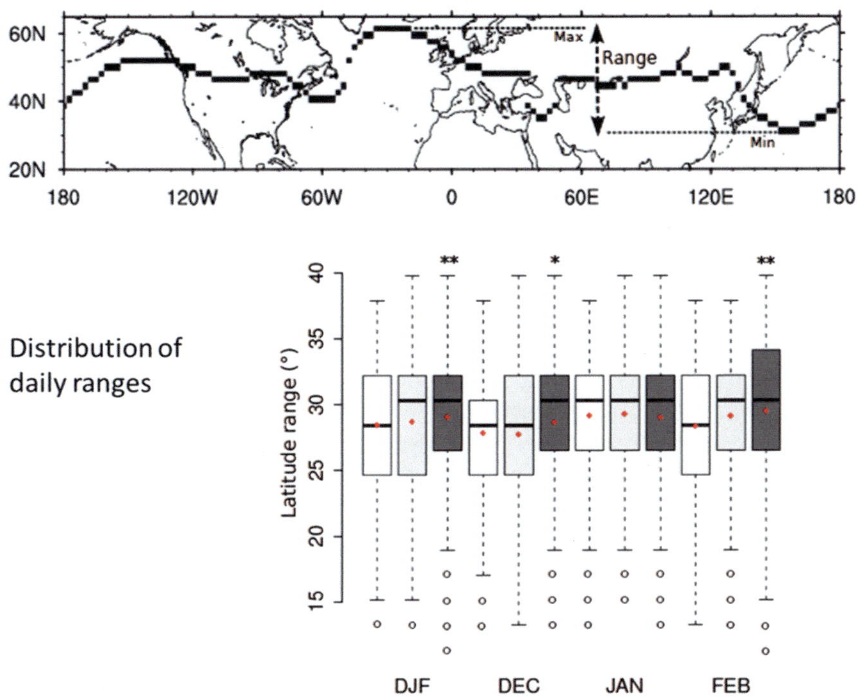

FIGURE 24 A weak but statistically significant increase in wave amplitude for 2090C. Top: Range of the 5400 m isoline of height on the 500 hPa surface for one day of February in the control. Bottom: Distribution of the 5400 m isopleth daily wave amplitude change in winter (DJF). White (control), light grey (2010C), dark grey (2090C). Red diamonds show the mean of the distribution, and asterisks indicate the significance level of the change of the mean in 2010C and 2090C, compared to the control value. SOURCE: Peings and Magnusdottir, 2014.

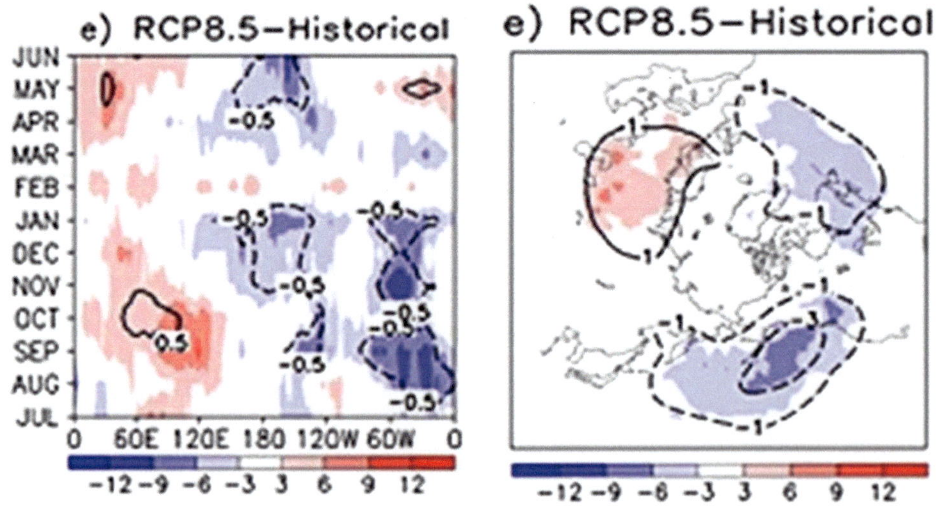

FIGURE 25 RCP8.5 minus historical multimodel mean. Left: Seasonal cycle of the Northern Hemisphere (NH) blocking frequency as a function of longitude. Units are days per month. Right: Climatology of NH annual-mean blocking frequency. Units are number of blocked days per year. SOURCE: Dunn-Sigouin and Son (2013).

Blocking

Barnes noted that some modeling studies find a decrease in high-latitude blocking events over the oceans, but an increase over Asia. The decreases are largest over the oceans in autumn and early winter. A couple of modeling studies find that decreases in blocking events are linked to changes in the jet stream (i.e. Barnes et al., 2012; de Vries et al., 2013). She noted that similar responses are found using other blocking identifying schemes (de Vries et al., 2013; Masato et al., 2013). Dunn-Sigouin and Son (2013) found RCP 8.5 integrations compared to historical[12] integrations, showing significant decreases in blocking frequency over both the North Pacific and North Atlantic regions, with slight increasing blocking frequency over western Russia (Figure 25).

Weather Patterns and Storm Tracks

Workshop participants discussed several modeling studies on changing weather patterns and storm tracks. Francis presented output from a CCSM4 RCP8.5 scenario that projected an increased ridging over continents associated with higher surface air temperatures. She noted that the ridging over the North Atlantic cools Western Europe. She also presented modeling evidence that projects that light and easterly winds will become more common at the expense of west winds for the 2090s. This will result in higher surface air temperatures in the mid-latitudes, which makes extremes more likely.

Tim Woollings, University of Oxford, discussed a set of experiments with one atmospheric model (HadGAM2) to investigate sources of spread in CMIP model projections of the Atlantic storm track response to climate change. This work focused on the late 21st century and showed that uncertainty in the magnitude of sea ice retreat is a large source of uncertainty in predicting the response of the storm track.

[12] Historical runs are 20th century climate integrations with all observed climate forcings.

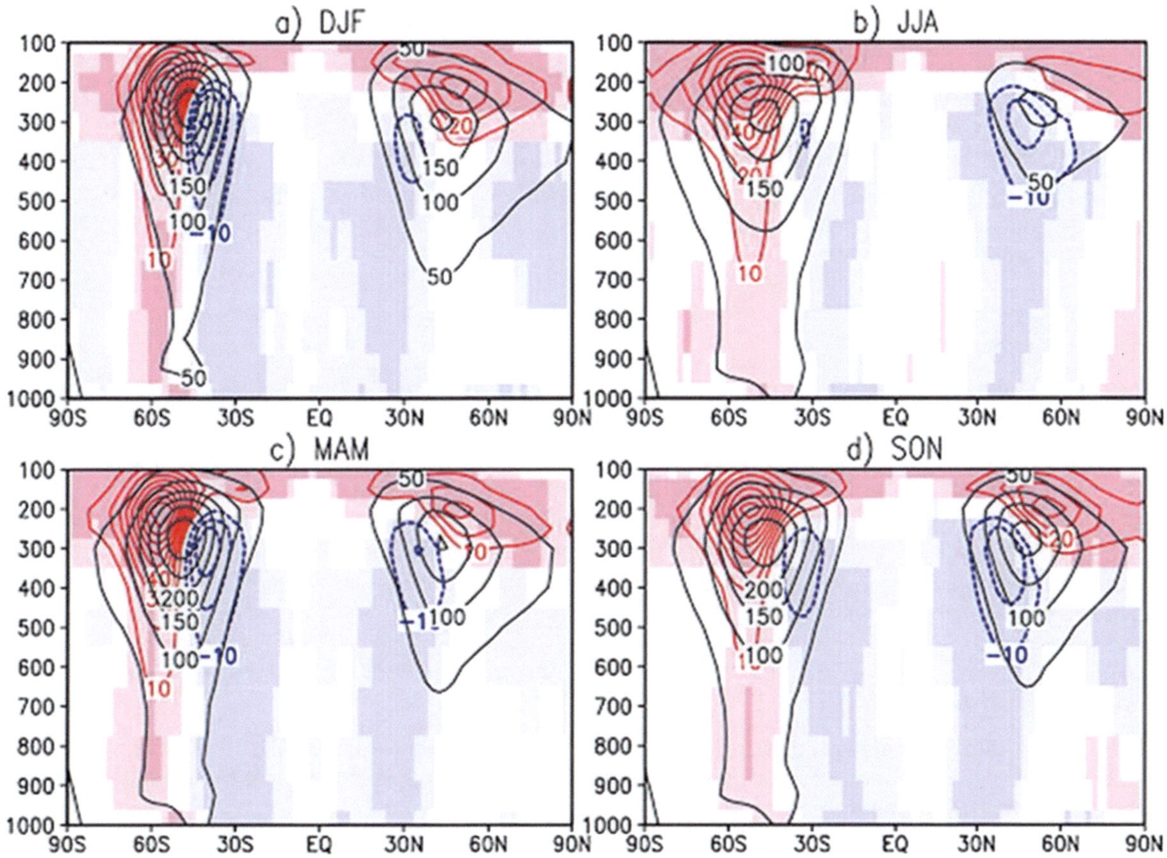

FIGURE 26 In the Northern Hemisphere (NH), the storm track is weakest during summer (JJA). Projected changes from 1980 to 1999 to 2081 to 2100 in zonal mean pressure-level variance statistics by CMIP5 multimodel ensemble based on RCP8.5 scenario, as a function of latitude and pressure. (a) DJF, (b) JJA, (c) MAM, (d) SON. Black contours indicate model climatology. Red and blue contours indicate projected changes. Shadings indicate regions over which 80 percent (light) or 100 percent (dark) of the models agree on the sign of the change. SOURCE: Chang et al., 2012.

Barnes presented a model study that projects that winter Northern Hemisphere upper-level storm tracks will weaken on equatorward flank by 2100 under RCP8.5; the storm track is weakest during summer (JJA) in the Northern Hemisphere (Figure 26; Chang et al., 2012).

Magnusdottir said that current Arctic sea-ice conditions (2010C) favor more intense cold extremes over mid-latitudes (mostly confined to the Asian sector; Peings and Magnusdottir 2014). With stronger sea-ice forcing (2090C), the intensity of cold extremes decreases everywhere north of 45 N because of the extension of the Arctic warm anomaly over northern continents. In 2090C (as in 2010C) cold extremes are more intense south of 45 N. Despite far stronger forcing in 2090C compared to 2010C, the intensity of cold extremes does not change significantly. Magnusdottir noted that this implies that there is a nonlinear relationship between sea-ice retreat and mid-latitude temperature. Liu et al. (2012) also found that a reduction in Arctic sea ice leads to more frequent cold weather outbreaks in the northern mid-latitudes.

Several additional modeling studies were briefly presented which highlight possible regional climate impacts when forced with a reduction in Arctic sea ice (Box 3).

BOX 3

Highlights of Other Modeling Studies Showing Regional Effects

Cold Winters in Northern Europe

Yang and Christensen (2012) found that cold early winter weather in northern Europe is associated with 20-40 percent reduction in Barents-Kara sea ice. Petoukhov and Semenov (2010) found that sea ice loss triples the probability of cold winter over large areas of the mid-latitudes including Europe.

Wet European Summers

There is modeling evidence that sea ice loss may lead to a southward shift of the summer jet stream over Europe, resulting in increased summer precipitation in the United Kingdom (Screen, 2013). Screen noted that the simulated increase is small and closely resembles the spatial pattern of precipitation anomalies in recent summers.

East Asian Monsoon

Wu et al. (2013) found that composites of high ice minus low ice lead to dry air over western Europe and far-east Asia and increased precipitation over west Asia.

Winter Mediterranean Rain

Grassi et al. (2013) utilized simulations that project that sea ice reduction will lead to stronger and more frequent extreme cold events over continental Europe and extreme precipitation events over the entire Mediterranean Basin. In particular, simulations suggest an increased risk of winter flooding in southern Italy, Greece, and the Iberian Peninsula.

Changes in Energy Transport Due to Arctic Amplification

Frierson discussed how poleward energy transport into the Arctic might change in the future (because of global warming). He noted that global circulation models indicate that *more* Arctic amplification is associated with *less* heat transport into the Arctic (Hwang et al., 2011). He presented modeling studies that show that energy actually transports toward the mid-latitudes (rather than toward the poles as we might expect) under Arctic amplification. He believes that this anti-correlation is due to the heat transport responding to temperature gradients. Arctic amplification causes weaker temperature gradients which leads to less dry static energy transport into the Arctic. More gradient results in more transport (i.e., transport is diffusive). This "diffusive framework" implies that Arctic warming should spread to lower latitudes (i.e., warmer Arctic leads to warmer mid-latitudes). Models with more Arctic warming have anomalous dry static energy transport southward, back toward the mid-latitudes. This occurs relatively independently of the storm track amplitude and location change within the model.

Arctic Sea Ice Predictability

Some workshop participants noted that Arctic sea ice predictions could eventually be used to assess the risk of temperature and precipitation anomalies and extreme weather events over northern continents. In her remarks, Holland discussed the potential to predict Arctic sea ice on seasonal timescales. She noted that seasonal sea ice prediction is an initial value problem (i.e., the predictability characteristics are largely dependent on the initial conditions). Potential predictability of sea ice decreases during spring and then is regained during summer and the following winter. Perfect model studies[13] consistently show that ocean heat content (temperature) leads to predictability in winter sea ice, in all cases. Summer sea ice predictability is associated with memory that resides in the ice itself—ice thickness. Therefore, there is reduced summer sea ice predictability in thin ice conditions. Model studies suggest limited predictability in sea ice area on 1- to 2-year timescales.

Holland said that realizing even a fraction of the potential predictability of sea ice requires an "adequate" observational network and "adequate" (and compatible) models. However, it is currently unclear what "adequate" means. We do not even know the spatial coverage and variables that are required in an "adequate" observational network. What model biases impact predictability, and what model complexity is required? However, Holland noted that some studies have shown some predictive skill (e.g. Sigmond et al., 2013; Wang et al., 2013).

Holland argued that the inherent limits of sea ice predictions coupled with the need for system improvements to realize predictability should be acknowledged as an important limitation with respect to the possible influence of sea ice loss on extreme weather.

Model Limitations: CMIP Biases

As highlighted in this workshop, CMIP5 model projections can be a useful tool for studying linkages between Arctic amplification and mid-latitude circulation. However, as Barnes noted in her remarks, they contain biases that should be considered. For example, models tend to place the Northern Hemisphere jet too far equatorward (no model is poleward of the observed North Atlantic jet in the annual-mean). Barnes also said that models tend to overestimate the seasonal cycle in the north Atlantic and storm track intensity is too weak (Zappa et al., 2014). It is also well-known that models underestimate observed high-latitude blocking frequencies. These underestimates are largest in the cool seasons. In general, increased horizontal and vertical resolution improves model blocking distributions (Anstey et al., 2013; Matsueda et al., 2009). Blocking biases in global circulation models can be reduced by correcting for mean circulation biases, but this does not remove all errors, especially at high latitudes, said Barnes.

Conclusion

The modeling studies discussed represent a fraction of the work that has been done to test possible mechanisms of the proposed linkages between Arctic amplification and mid-latitude weather. Because the modeling studies are largely divergent, many participants believe research in this area should continue. In particular, many participants said that the modeling community should organize careful model intercomparisons and model sensitivity studies with similar boundary and/or initial conditions to allow for a more systematic review and comparison of the results. Peings and Magnusdottir (2014) provide a detailed comparison of some of the recent modeling work.

Although the workshop presentations highlighted several model limitations and biases, many participants asserted that models are the best tools available to complement observational studies, and to test robustness of the proposed mechanisms.

---

[13] A perfect model study is an idealized exercise where an ensemble of forecasts from a single model are compared to one another. The ensemble is typically initialized from some point in time from the same model from an earlier run. The perfect model exercise is not initialized with observations or any attempt to assimilate observations into the model, and hence the initial state is a state of the model, not nature. The evaluation of the forecast therefore avoids any inaccuracy in knowledge of the observed initial state, and it avoids model errors in reproducing nature.

ACID TEST TO ASSESS PROPOSED LINKAGES

It was clear from the discussion that research into the linkages is still emerging and will continue in the future. Screen proposed using an "ACID test" as a framework for assessing confidence in proposed linkages: Attribution, Corroboration, Informed by mechanistic understanding, Detection. The first step is to ask whether the proposed link is attributable to sea ice loss (or another forcing). It is important to note that correlation does not necessarily imply causation. Screen stressed that it is difficult (if not impossible) to identify cause and effect in the real world. Carefully designed simulations are key for determining attribution (e.g., decreased mid-tropospheric north-south temperature gradient cannot be attributed to sea ice loss alone). It is also important to differentiate between sea ice loss and Arctic amplification (which has multiple causes). Also, observations do not tell us much about cause and effect, but models can.

The next step is to determine if the proposed link is corroborated by multiple lines of evidence. Is there agreement among multiple studies? Is there agreement between models? Different models with the same forcing can actually give opposite responses. All of the different factors that may affect the model outcome must be considered. Is there agreement among metrics?

Third, is the proposed linkage informed by mechanistic understanding? Confidence in the existence of a particular linkage is clearly strengthened by an in-depth understanding of the mechanisms. Screen acknowledged that much work is left to be done in this area. Do we understand the mechanisms driving the linkages?

The final step in Screen's ACID test is to ask whether the link is detectable in the real world? Is the signal large enough to be detected over intrinsic/natural variability? Is the signal strong enough to detect over other forcings? Some types of responses are easier to detect than others (see Figure 4).

Screen believes that few, if any, of the proposed linkages pass the ACID test and that caution is needed in linking sea ice loss to recent events and trends. More research is required to make more confident statements. Other participants noted that the ACID test approach is sound, but, given the limitations of available information, there are inherent limitations to the analyses that can be conducted.

FUTURE NEEDS AND OPPORTUNITIES

Workshop participants divided into three breakout groups to discuss future needs and opportunities in (a) observations, (b) models, and (c) a big picture context. A summary of the discussions in each of these three breakout groups follows.

Observations

One theme that emerged was the need to more effectively utilize existing observational capabilities. For example, there was a sense that research aircraft surveys targeted to key seasons and processes are underutilized. Participants also noted that a vast amount of data is never analyzed. Department of Defense products such as Visible and Infrared Scanner (VIRS) and National Ice Center (NIC) operational products could be useful and are not saved unless there is a perceived need for them.

Furthermore, making observations from data available more quickly would allow researchers enough time to analyze them.

Individual participants identified other specific observational and data needs, including

- quantifying uncertainties in data products to clarify trends (e.g., sea ice concentration),
- reprocessing historical snow products,
- utilizing a more robust analysis of hourly temperature data to accurately represent the diurnal cycle in the Arctic,
- deploying field programs that target extreme storm events (especially moisture measurements to clarify role of latent heating),
- obtaining measurements that target:
    - clouds and surface fluxes (energy/radiation, moisture, momentum), especially over an increasingly open Arctic, from Arctic buoys (especially under high-wind conditions) and autonomous remote surface observations;
    - atmosphere in the Arctic to better understand vertical temperature structure, stratification (e.g. using dropsondes from unmanned aircraft) and changes in cyclone/storm tracks; and
    - the Arctic Ocean to improve seasonal predictions, and
- deploying process driven studies, for example, to understand the processes that drive sea ice.

Modeling

Another theme that emerged from the workshop discussion was the importance of model sensitivity studies and model inter-comparison studies to provide clues on the mechanisms of potential Arctic linkages, as well as for detection and attribution. These studies require models that are consistently developed with similar boundary conditions (e.g., sea ice extent and thickness).

Individual workshop participants identified other needs related to modeling:

- Conducting idealized modeling studies to:
    - understand how Arctic amplification and the atmospheric circulation response are coupled (e.g. baroclinic vs. barotropic, seasonality);
    - understand future wave propagation with respect to Arctic change and the possible NAO-type response to sea ice loss; and
    - capture seasonality and longitudinal variations (by trying perturbed physics runs, and not just fixed boundary conditions).
- Conducting attribution studies to understand the proportion of Arctic amplification that is due to sea ice vs. the proportion that is due to processes outside of the Arctic.
- Identifying the resolution that is required to correctly model fluxes (including horizontal and vertical atmosphere and boundary conditions).
- Applying the sophisticated methods and diagnostics that have been used in the meteorological community (e.g., see John Gyakum's presentation) to this issue.
- Developing higher resolution models to study processes in the planetary boundary layer (below 850 hPa or roughly 1,500 m above sea level).
- Developing convection-permitting global climate models that resolve to better simulate vertical wind shear.

Big Picture Context

The third breakout group discussed the key points that emerged from the workshop discussions. Individuals in the group identified the following key messages:

- Participants still disagree about the influence of recent Arctic warming on mid-latitude weather patterns.
- Research on the Arctic influence on mid-latitudes is at an early stage compared to influences such as tropical, mid-latitude SST, or the stratosphere. The recent proposed influences are best viewed as hypotheses.
- It is important to retain a global perspective by considering the influence of the Arctic in the context of forcing from other components of the climate system, such as the tropics and the stratosphere. Changes in ocean circulation may be particularly important in Arctic change and its influence on the mid-latitudes. A key limitation is lack of knowledge of the Arctic Ocean energy budget.
- Attention has so far focused on changes in the meridional temperature gradient. Some consideration could be given to other mechanisms, such as zonal temperature gradients, static stability and moisture changes. The possibility of distinct regional responses could also be considered (e.g., east and west Atlantic).
- Application of more sophisticated methods of analysis to determine whether mid-latitude wave activity is changing would be useful. A long history of dynamical analysis of wave activity can be exploited in this field. Additionally, the link between Arctic warming and weakened westerlies is not clear, because of transient eddy feedbacks, which often act to shift the westerlies instead.
- More effort should be directed to understanding the synoptic aspects of possible linkages, considering the variety of synoptic weather systems active in the region, and to bridging the gap between case study analysis and filtered variance approaches. This topic could be further developed through collaboration between weather and climate scientists.
- Blocking is important for weather impacts and was a recurring feature of the workshop discussion, yet it is often interpreted differently by different people or analyses. Improving our understanding of the range of blocking systems and how they may be influenced by forcing is critical.
- It is possible that recent Arctic changes have pushed the atmosphere into a new state with different variability. The strong Arctic forcing has emerged only in the past few years, and development of new methods and approaches may be required to test or account for it. For example, it might be useful to adopt a probabilistic approach to determine whether there are changes in the probability of extreme events.
- Understanding of natural variability should be improved to determine whether changes are taking place. Reanalyses of the atmosphere are generally trustworthy for synoptic and large-scale features, so relatively long records are available. Work to extend sea ice records back in time is also useful. Ocean observations are seriously limited.
- Combining observations with a hierarchy of models is key to making progress on this issue. The detection and attribution framework provides one formal approach.
- Arctic influences could provide additional sources of skill for seasonal forecasting through model improvements or statistical methods.
- A recurring issue is how to respond to the public when asked questions such as "I heard that sea ice loss caused the wet summer, why don't you include that in your forecasts?"
- A large, policy-relevant research program on these issues would bring the subject to maturity in the same way that has occurred for similar subjects (e.g., stratospheric, ocean influences). The program could focus on specific topics, for example, identifying something we are trying to explain, or by being provocative and picking on the weak links of chains of arguments.

- Although many challenges exist, several new opportunities will make significant progress in this area possible in the near future. These include (a) improvements in atmospheric models, some of which can now simulate blocking well, (b) similar improvements in ocean and sea ice models, (c) longer records back into the past of some sea ice quantities and atmospheric circulation, (d) emergence of reasonably strong sea ice forcing in recent years, and (e) strong interest in the topic from both public and governments.

Overall, many participants thought that, given the far-reaching societal impacts of changes in our climate system and the speed at which the Arctic sea ice cover is disappearing, it is imperative to advance understanding of this issue and to assess what it may mean for the frequency, severity, and persistence of severe weather events in the coming years. The workshop discussions made it clear that progress on this front requires more observational work and modeling experiments to place Arctic linkages within a larger context of the other factors that affect mid-latitude weather.

References

Anstey, J. A., P. Davini, L. J. Gray, T. J. Woollings, N. Butchart, C. Cagnazzo, B. Christiansen, S. C. Hardiman, S. M. Osprey, and S. T. Yang. 2013. Multi-model analysis of Northern Hemisphere winter blocking: Model biases and the role of resolution. Journal of Geophysical Research-Atmospheres 118(10):3956-3971.

Baldwin, M. P. and T. J. Dunkerton. 2001. Stratospheric harbingers of anomalous weather regimes. Science 294(5542):581-584.

Barnes, E. A. 2013. Revisiting the evidence linking Arctic amplification to extreme weather in midlatitudes. Geophysical Research Letters 40:1-6.

Barnes, E. A., J. Slingo, and T. Woollings. 2012. A methodology for the comparison of blocking climatologies across indices, models and climate scenarios. Climate Dynamics 38(11-12):2467-2481.

Benedict, J. J., S. Lee, and S. B. Feldstein. 2004. Synoptic view of the North Atlantic Oscillation. Journal of the Atmospheric Sciences 61(2):121-144.

Berrisford, P., B. J. Hoskins, and E. Tyrlis. 2007. Blocking and Rossby wave breaking on the dynamical tropopause in the Southern Hemisphere. Journal of the Atmospheric Sciences 64(8):2881-2898.

Black, E., M. Blackburn, G. Harrison, B. Hoskins, and J. Methven. 2004. Factors contributing to the summer 2003 European heatwave. Weather 59(8):217-223.

Ceppi, P., Y. T. Hwang, X. J. Liu, D. M. W. Frierson, and D. L. Hartmann. 2013. The relationship between the ITCZ and the Southern Hemispheric eddy-driven jet. Journal of Geophysical Research-Atmospheres 118(11):5136-5146.

Chang, E. K. M., Y. Guo, and X. Xia. 2012. CMIP5 multi-model ensemble projection of storm track change under global warming. Journal of Geophysical Research-Atmospheres 117.

Chen, G. and I. M. Held. 2007. Phase speed spectra and the recent poleward shift of Southern Hemisphere surface westerlies. Geophysical Research Letters 34(21).

Chen, G., J. Lu, and D. M. W. Frierson. 2008. Phase Speed Spectra and the Latitude of Surface Westerlies: Interannual Variability and Global Warming Trend. Journal of Climate 21(22):5942-5959.

Chiang, J. C. H. and C. M. Bitz. 2005. Influence of high latitude ice cover on the marine Intertropical Convergence Zone. Climate Dynamics 25(5):477-496.

Cohen, J., M. Barlow, P. J. Kushner, and K. Saito. 2007. Stratosphere-troposphere coupling and links with Eurasian land surface variability. Journal of Climate 20(21):5335-5343.

Cohen, J. and D. Entekhabi. 1999. Eurasian snow cover variability and Northern Hemisphere climate predictability. Geophysical Research Letters 26(3):345-348.

Cohen, J. L., J. C. Furtado, M. A. Barlow, V. A. Alexeev, and J. E. Cherry. 2012. Arctic warming, increasing snow cover and widespread boreal winter cooling. Environmental Research Letters 7(1).

Coumou, D. and S. Rahmstorf. 2012. A decade of weather extremes. Nature Climate Change 2(7):491-496.

Croci-Maspoli, M., C. Schwierz, and H. C. Davies. 2007. A multifaceted climatology of atmospheric blocking and its recent linear trend. Journal of Climate 20(4):633-649.

Davini, P., C. Cagnazzo, P. G. Fogli, E. Manzini, S. Gualdi, and A. Navarra. 2013. European blocking and Atlantic jet stream variability in the NCEP/NCAR reanalysis and the CMCC-CMS climate model. Climate Dynamics doi: 10.1007/s00382-013-1873-y.

Davini, P., C. Cagnazzo, S. Gualdi, and A. Navarra. 2012. Bidimensional Diagnostics, Variability, and Trends of Northern Hemisphere Blocking. Journal of Climate 25(19):6496-6509.

de Vries, H., T. Woollings, J. Anstey, R. J. Haarsma, and W. Hazeleger. 2013. Atmospheric blocking and its relation to jet changes in a future climate. Climate Dynamics 41(9-10):2643-2654.

Dole, R., M. Hoerling, J. Perlwitz, J. Eischeid, P. Pegion, T. Zhang, X. W. Quan, T. Y. Xu, and D. Murray. 2011. Was there a basis for anticipating the 2010 Russian heat wave? Geophysical Research Letters 38.

Dunn-Sigouin, E. and S. W. Son. 2013. Northern Hemisphere blocking frequency and duration in the CMIP5 models. Journal of Geophysical Research-Atmospheres 118(3):1179-1188.

Francis, J. A. and S. J. Vavrus. 2012. Evidence linking Arctic amplification to extreme weather in mid-latitudes. Geophysical Research Letters 39.

Frierson, D. M. W., J. Lu, and G. Chen. 2007. Width of the Hadley cell in simple and comprehensive general circulation models. Geophysical Research Letters 34(18).

Ghatak, D., C. Deser, A. Frei, G. Gong, A. Phillips, D. A. Robinson, and J. Stroeve. 2012. Simulated Siberian snow cover response to observed Arctic sea ice loss, 1979-2008. Journal of Geophysical Research-Atmospheres 117.

Gong, G., D. Entekhabi, and J. Cohen. 2002. A large-ensemble model study of the wintertime AO-NAO and the role of interannual snow perturbations. Journal of Climate 15(23):3488-3499.

Grassi, B., G. Redaelli, and G. Visconti. 2013. Arctic Sea Ice Reduction and Extreme Climate Events over the Mediterranean Region. Journal of Climate 26(24):10101-10110.

Haarsma, R. J., W. Hazeleger, C. Severijns, H. de Vries, A. Sterl, R. Bintanja, G. J. van Oldenborgh, and H. W. van den Brink. 2013a. More hurricanes to hit western Europe due to global warming. Geophysical Research Letters 40(9):1783-1788.

Haarsma, R. J., F. Selten, and G. J. van Oldenborgh. 2013b. Anthropogenic changes of the thermal and zonal flow structure over Western Europe and Eastern North Atlantic in CMIP3 and CMIP5 models. Climate Dynamics 41(9-10):2577-2588.

Harvey, B. J., L. C. Shaffrey, and T. J. Woollings. 2013. Large-scale temperature gradients and the extratropical storm track responses in CMIP5. 13th EMS / 11th ECAM. EMS Annual Meeting Abstracts 10:EMS2013-2508.

Hurrell, J. W. 1995. Decadal Trends in the North-Atlantic Oscillation - Regional Temperatures and Precipitation. Science 269(5224):676-679.

Hwang, Y. T., D. M. W. Frierson, and J. E. Kay. 2011. Coupling between Arctic feedbacks and changes in poleward energy transport. Geophysical Research Letters 38.

Jahn, A. and M. M. Holland. 2013. Implications of Arctic sea ice changes for North Atlantic deep convection and the meridional overturning circulation in CCSM4-CMIP5 simulations. Geophysical Research Letters 40(6):1206-1211.

Kidston, J. and E. P. Gerber. 2010. Intermodel variability of the poleward shift of the austral jet stream in the CMIP3 integrations linked to biases in 20th century climatology. Geophysical Research Letters 37.

Kug, J. S., D. H. Choi, F. F. Jin, W. T. Kwon, and H. L. Ren. 2010. Role of synoptic eddy feedback on polar climate responses to the anthropogenic forcing. Geophysical Research Letters 37.

Lee, S., S. Feldstein, D. Pollard, and T. White. 2011a. Do Planetary Wave Dynamics Contribute to Equable Climates? Journal of Climate 24(9):2391-2404.

Lee, S., T. T. Gong, N. Johnson, S. B. Feldstein, and D. Pollard. 2011b. On the Possible Link between Tropical Convection and the Northern Hemisphere Arctic Surface Air Temperature Change between 1958 and 2001. Journal of Climate 24(16):4350-4367.

Liu, J. P., J. A. Curry, H. J. Wang, M. R. Song, and R. M. Horton. 2012. Impact of declining Arctic sea ice on winter snowfall. Proceedings of the National Academy of Sciences of the United States of America 109(11):4074-4079.

Lorenz, D. J. and E. T. DeWeaver. 2007. Tropopause height and zonal wind response to global warming in the IPCC scenario integrations. Journal of Geophysical Research-Atmospheres 112(D10).

Lu, J., G. Chen, and D. M. W. Frierson. 2010. The Position of the Midlatitude Storm Track and Eddy-Driven Westerlies in Aquaplanet AGCMs. Presented at American Geophysical Union Fall Meeting 2010, San Francisco, CA.

Masato, G., B. J. Hoskins, and T. Woollings. 2013. Winter and Summer Northern Hemisphere Blocking in CMIP5 Models. Journal of Climate 26(18):7044-7059.

Massonnet, F., T. Fichefet, H. Goosse, C. M. Bitz, G. Philippon-Berthier, M. M. Holland, and P. Y. Barriat. 2012. Constraining projections of summer Arctic sea ice. Cryosphere 6(6):1383-1394.

Matsueda, M., R. Mizuta, and S. Kusunoki. 2009. Future change in wintertime atmospheric blocking simulated using a 20-km-mesh atmospheric global circulation model. Journal of Geophysical Research-Atmospheres 114.

Mizuta, R. 2012. Intensification of extratropical cyclones associated with the polar jet change in the CMIP5 global warming projections. Geophysical Research Letters 39.

NRC (National Research Council). 2010. Assessment of Intraseasonal to Interannual Climate Prediction and Predictability. Washington, DC: National Academies Press.

Overland, J. E. and M. Y. Wang. 2010. Large-scale atmospheric circulation changes are associated with the recent loss of Arctic sea ice. Tellus Series a-Dynamic Meteorology and Oceanography 62(1):1-9.

Overland, J. E., K. R. Wood, and M. Y. Wang. 2011. Warm Arctic-cold continents: climate impacts of the newly open Arctic Sea. Polar Research 30.

Peings, Y. and G. Magnusdottir. 2014. Response of the Wintertime Northern Hemisphere Atmospheric Circulation to Current and Projected Arctic Sea Ice Decline: A Numerical Study with CAM5. Journal of Climate 27:244-264.

Perlwitz, J. and N. Harnik. 2004. Downward coupling between the stratosphere and troposphere: The relative roles of wave and zonal mean processes. Journal of Climate 17(24):4902-4909.

Petoukhov, V. and V. A. Semenov. 2010. A link between reduced Barents-Kara sea ice and cold winter extremes over northern continents. Journal of Geophysical Research-Atmospheres 115.

Pithan, F. and T. Mauritsen. 2014. Arctic amplification dominated by temperature feedbacks in contemporary climate models. Nature Geoscience 7:181-184.

Polvani, L. M. and D. W. Waugh. 2004. Upward wave activity flux as a precursor to extreme stratospheric events and subsequent anomalous surface weather regimes. Journal of Climate 17(18):3548-3554.

Riddle, E. E., A. H. Butler, J. C. Furtado, J. L. Cohen, and A. Kumar. 2013. CFSv2 ensemble prediction of the wintertime Arctic Oscillation. Climate Dynamics 41(3-4):1099-1116.

Screen, J. A. 2013. Influence of Arctic sea ice on European summer precipitation. Environmental Research Letters 8(4):044015.

Screen, J. A., C. Deser, I. Simmonds, and R. Tomas. 2013. Atmospheric impacts of Arctic sea-ice loss, 1979-2009: separating forced change from atmospheric internal variability. Climate Dynamics doi:10.1007/s00382-013-1830-9.

Screen, J. A. and I. Simmonds. 2013. Exploring links between Arctic amplification and mid-latitude weather. Geophysical Research Letters 40(5):959-964.

Shaw, T. A. and J. Perlwitz. 2013. The Life Cycle of Northern Hemisphere Downward Wave Coupling between the Stratosphere and Troposphere. Journal of Climate 26(5):1745-1763.

Shindell, D., D. Rind, N. Balachandran, J. Lean, and P. Lonergan. 1999. Solar cycle variability, ozone, and climate. Science 284(5412):305-308.

Sigmond, M., J. C. Fyfe, G. M. Flato, V. V. Kharin, and W. J. Merryfield. 2013. Seasonal forecast skill of Arctic sea ice area in a dynamical forecast system. Geophysical Research Letters 40(3):529-534.

Sigmond, M., J. F. Scinocca, and P. J. Kushner. 2008. Impact of the stratosphere on tropospheric climate change. Geophysical Research Letters 35(12).

Stroeve, J. C., M. C. Serreze, M. M. Holland, J. E. Kay, J. Malanik, and A. P. Barrett. 2012. The Arctic's rapidly shrinking sea ice cover: a research synthesis. Climatic Change 110(3-4):1005-1027.

Strong, C. and G. Magnusdottir. 2008. Tropospheric Rossby wave breaking and the NAO/NAM. Journal of the Atmospheric Sciences 65(9):2861-2876.

Tang, Q. H., X. J. Zhang, X. H. Yang, and J. A. Francis. 2013. Cold winter extremes in northern continents linked to Arctic sea ice loss. Environmental Research Letters 8(1).

Thompson, D. W. J. and J. M. Wallace. 2000. Annular modes in the extratropical circulation. Part I: Month-to-month variability. Journal of Climate 13(5):1000-1016.

van Vuuren, D. P., J. Edmonds, M. Kainuma, K. Riahi, A. Thomson, K. Hibbard, G. C. Hurtt, T. Kram, V. Krey, J. F. Lamarque, T. Masui, M. Meinshausen, N. Nakicenovic, S. J. Smith, and S. K. Rose. 2011. The representative concentration pathways: an overview. Climatic Change 109(1-2):5-31.

Wang, W. Q., M. Y. Chen, and A. Kumar. 2013. Seasonal Prediction of Arctic Sea Ice Extent from a Coupled Dynamical Forecast System. Monthly Weather Review 141(4):1375-1394.

Winton, M. 2006. Amplified Arctic climate change: What does surface albedo feedback have to do with it? Geophysical Research Letters 33(3).

WMO (World Meteorological Organization). 2011. 2011: World's 10th warmest year, warmest year with La Niña on record, second-lowest Arctic sea ice extent. Provisional Statement on the Status of the Global Climate. World Meteorological Organization, Geneva.

Woollings, T., A. Hannachi, and B. Hoskins. 2010. Variability of the North Atlantic eddy-driven jet stream. Quarterly Journal of the Royal Meteorological Society 136(649):856-868.

Woollings, T., B. Harvey, M. Zahn, and L. Shaffrey. 2012. On the role of the ocean in projected atmospheric stability changes in the Atlantic polar low region. Geophysical Research Letters 39.

Woollings, T., B. Hoskins, M. Blackburn, and P. Berrisford. 2008. A new Rossby wave-breaking interpretation of the North Atlantic Oscillation. Journal of the Atmospheric Sciences 65(2):609-626.

Wu, B. Y., R. H. Zhang, R. D'Arrigo, and J. Z. Su. 2013. On the Relationship between Winter Sea Ice and Summer Atmospheric Circulation over Eurasia. Journal of Climate 26(15):5523-5536.

Yang, S. T. and J. H. Christensen. 2012. Arctic sea ice reduction and European cold winters in CMIP5 climate change experiments. Geophysical Research Letters 39.

Yoo, C., S. Feldstein, and S. Lee. 2011. The impact of the Madden-Julian Oscillation trend on the Arctic amplification of surface air temperature during the 1979-2008 boreal winter. Geophysical Research Letters 38.

Yoo, C., S. Lee, and S. Feldstein. 2012a. The impact of the Madden-Julian Oscillation trend on the inter-decadal Antarctic warming during the 1979-2008 austral winter. Atmospheric Science Letters 13:194-199.

Yoo, C., S. Lee, and S. B. Feldstein. 2012b. Arctic Response to an MJO-Like Tropical Heating in an Idealized GCM. Journal of the Atmospheric Sciences 69(8):2379-2393.

Zappa, G., G. Masato, L. Shaffrey, T. Woollings, and K. Hodges. 2014. Linking Northern Hemisphere blocking and storm track biases in the CMIP5 climate models. Geophysical Research Letters 41(1):135-139.

Appendix A
Speaker Abstracts

SESSION #1
BIG PICTURE CONTEXT: THE ROLE OF ARCTIC SEA ICE LOSS VS. OTHER FORCING FACTORS

Introduction to Discussion of Big Picture Context

This talk raised some general points about the issues of distinguishing Arctic sea ice loss from other forcings in considering mid-latitude weather and extremes in preparation for a broad brainstorming session in the first morning of the workshop. We started with the idea that Arctic sea ice loss is in a sense the middle of a story that really begins with amplified Arctic change forced by greenhouse warming. In this bigger picture, Arctic sea ice, atmosphere, and ocean processes are strongly coupled and are all relevant controls on mid-latitude weather. We encouraged participants to think systematically about how Arctic changes might directly influence the mid-latitudes, and about how changes in mid-latitudes might be affected by the Arctic through feedbacks. We also pointed out that any change that projects onto modes of variability such as the NAM/AO may simultaneously affect the Arctic and mid-latitudes, making physical attribution more difficult. Finally, we proposed that efforts should be made to be precise about the timescale and seasonality of any proposed linkages, to think about the quantitative evidence for our current understanding (such as the ACID test of Screen), and to consider critical research gaps (observational/modelling/theoretical/impacts-related) in addressing the important question of Arctic influence on mid-latitude weather and extremes.

The Role of Arctic Sea Ice Loss
James Screen, University of Exeter

Many linkages have been proposed between Arctic climate change and mid-latitude weather. These have received substantial media coverage. In a recent New Hampshire poll, 60 percent of respondents answered that they thought that future rapid Arctic warming would have major effects on the weather where they live. This talk reviewed published scientific evidence for proposed Arctic-mid-latitude linkages, including between Arctic sea ice loss and cold mid-latitude winters, and the evidence for, and against, changing planetary wave amplitudes in response to greater warming in the Arctic than at lower latitudes. A framework for assessing confidence in proposed linkages was introduced, the "ACID" test. It was argued that to have high confidence in a linkage it should be Attributable to Arctic forcing, Corroborated by multiple lines of evidence, Informed by mechanistic understanding and Detectable in the real world. It was argued that few, if any, of the proposed linkages currently pass the ACID test. A number of key uncertainties were highlighted, including disagreement between the atmospheric responses to sea ice loss in different modeling studies. Also, it discussed how responses in some variables and locations are easier to detect (i.e. are stronger) than others. Pertinently, mid-latitude atmospheric responses are hard to detect due to remoteness from the sea-ice forcing (changes in surface fluxes) and large atmospheric internal variability. Thus, a degree of caution is required when linking sea ice loss to recent extreme weather events and possible trends therein. In short, the science behind Arctic-mid-latitude linkages is still in its infancy and more research is required to make more confident statements.

Present-Day and Future Projections of Mid-latitude Variability
Elizabeth Barnes, Colorado State University

Mid-latitude atmospheric variability is composed of complex dynamical interactions between circulations of a range of spatial and temporal scales. For example, the low-frequency variability of the Northern Hemisphere jet-stream is tightly coupled to the presence and absence of high-frequency patterns such as blocking anticyclones and Rossby wave-breaking.

It has been hypothesized that Arctic amplification has increased the variability of the jet-stream and the frequency of blocking over the North Atlantic/North America region, however, no statistically significant increases in blocking frequency have been observed. It is suggested that given the large internal variability of the jet-stream and blocking, any potential effect of Arctic amplification on mid-latitude weather is unlikely to be detectable with current observations.

Looking to the future, the model simulations performed for CMIP5 (the Coupled Model Intercomparison Project, phase 5) offer mixed evidence of a link between Arctic amplification and mid-latitude circulation patterns. Blocking frequencies are projected to decrease over the oceans with climate change, opposite to the increase hypothesized to go with Arctic amplification. However, significant correlations between atmospheric wave properties and Arctic temperatures are shown. Idealized model experiments are highlighted as a useful way forward to test proposed dynamical mechanisms linking Artic amplification and mid-latitude variability.

Increasing greenhouse gas concentrations are projected to influence not just the Arctic climate, but the climate at other latitudes as well. Thus, a goal of future research should be to assess the relative importance of Arctic change compared to these other regions in determining future mid-latitude weather and its extremes.

The Role of the Stratosphere and Seasonal Snow Cover
Paul J. Kushner, University of Toronto

This talk discussed two linkages that figure prominently in discussions of linkages between Arctic and the mid-latitudes: 1) coupling between the polar stratosphere and the extratropical troposphere, and 2) the related coupling between large-scale snow cover anomalies and circulation. Although stratospheric variability is mainly controlled by planetary waves propagating upward from the troposphere, the last 15 years have seen a large number of papers dealing with "stratospheric influence", which is a back effect of stratospheric variations on the tropospheric circulation. This talk identified two dynamical pathways for this influence on intraseasonal timescales, one involving wave mean flow interactions and the Northern Annular Mode/Arctic Oscillation (NAM/AO), as initially proposed by Baldwin and Dunkerton; and another involving wave reflection and the generation of regional circulation anomalies, as described by Perlwitz, Harnik, and Shaw. In these linkages, the role of the stratosphere is not always direct. In particular, stratospheric anomalies may have robust tropospheric precursors from sea ice, snow, tropical forcing, etc.. An example of this is the effect of tropical forcing on the Arctic via the polar stratosphere, which is particularly sensitive to the structure of the background stationary wave field through a linear interference effect (papers by Fletcher and Kushner).

Another example of tropospherically forced signals that involve the stratosphere as an intermediary is the proposed role of October Eurasian snow cover in Eurasia in driving wintertime NAM/AO anomalies, as described by Cohen and collaborators. The snow-NAM connection has been found to operate in interannual variability and provide a useful tool for seasonal forecasting. For long-term trends, connections between September sea ice loss, October Eurasian snow increase, and NAM anomalies have been drawn. The talk discussed the role of observational uncertainty and decadal variability in making it challenging to obtain a robust understanding of snow-NAM linkages.

3-Minute Open Session Presentations: Big Picture

Why has the Arctic Warmed?
Martin Hoerling, NOAA

A statement from Francis and Vavrus (2012) was presented: "The differential warming of the Arctic relative to mid- latitudes is the key linking Arctic amplification with patterns favoring persistent weather conditions in mid-latitudes." This was clarified to column warming. Conjectures from the same paper were also provided: observed integrated 1000-500mb Arctic warming is due to Arctic sea ice loss," and "observed weaker poleward thickness gradients are due to Arctic sea ice loss." A model was run both with and without sea ice to find the net effect of the ice. The estimated observed 1000-500hPa Arctic warming found was inconsistent with model internal atmospheric variability alone, was largely forced, but inconsistent with effects of sea ice loss alone, and thus could be largely due to natural decadal sea surface temperature (SST) forcing.

Arctic Contribution to Future Storm Track Uncertainty
Tim Woollings, University of Oxford

Two strands of relevant work were summarized. The first of these is a set of experiments with one atmospheric model (HadGAM2) to investigate sources of spread in CMIP model projections of the Atlantic storm track response to climate change. This work focuses on the late 21st century and shows that uncertainty in the magnitude of sea ice retreat is a large source of uncertainty in predicting the response of the storm track. The second strand of work looks at future changes in temperature variability. Horizontal temperature gradients are projected to weaken in the future, partly due to Arctic Amplification, and this may result in a reduction in temperature variability in the mid-latitudes in winter, since the effect of wind blowing across temperature gradients will lead to weaker temperature anomalies.

Open Panel Synopsis
John Walsh, University of Alaska, Fairbanks

This Open Panel presentation included two items. The first was a summary of the upcoming National Climate Assessment's (2013) message on linkages between Arctic sea ice loss and mid-latitude circulation. This message highlights the Francis and Vavrus (2012) finding that decreases in subpolar westerlies are correlated with Arctic sea ice loss, but cautions that such conclusions are dependent on the metric used for wave activity and are limited by the small sample size. The second item was a figure showing that 500 hPa zonal wind speeds averaged over 30-80ºN for the months of October-December have indeed weakened from 1979-present, consistent with summer sea ice loss. However, an extension of the time series back to 1948 shows an increase of zonal winds from the 1950s through the late 1970s. There is no known corresponding trend of (increasing) sea ice cover during the summer/autumn period in these decades.

The Other High Latitude Boundary Forcing: Snow
Judah Cohen, Atmospheric and Environmental Research

Though the dramatic reduction in sea ice and its impact on mid latitude weather especially extreme events has received much of the attention, Siberian snow cover has also been shown to influence mid-latitude winter weather. Comparison of correlations between October Siberian snow cover (as measured by the snow advanced index or SAI) and September Arctic sea ice extent and the winter Arctic Oscillation, the dominant mode of variability for the extratopical Northern Hemisphere, shows that the SAI is much more highly correlated with the winter AO than sea ice. An operational statistical model, which uses the SAI as its main predictor, correctly predicted the cold winter in 2013 across Northern Eurasia and the United States (pattern correlation between predicted and observed temperatures of 0.65). Furthermore the statistical model greatly outperformed a suite of dynamical models, which all predicted warm temperatures. The excellent forecast is consistent with other recent accurate forecasts.

Importance of Arctic Linkages for Policy and Communication
Benjamin J. DeAngelo, U.S. Environmental Protection Agency (EPA)

This submission offers a client perspective. There are two areas where EPA is engaged where this is relevant.

The first involves communication. EPA produces its periodic "Climate Change Indicators in the United States" (http://www.epa.gov/climatechange/science/indicators/index.html), which includes loss of Arctic sea ice as one of 26 indicators. Loss of Arctic sea ice already serves as an iconic communication tool; it can indicate the rate of climate change as well as 'how we're doing' in terms of GHG trajectories. Linkages between Arctic sea-ice loss and changes in weather patterns across the U.S. are an additional element that could be added to the Arctic storyline for communication purposes.

The second involves EPA's work under the Arctic Council—a forum of eight nations holding Arctic territory—to address short-lived climate forcers, primarily black carbon. The Arctic Council is working towards some sort of "arrangement" on short-lived climate forcers. Here too, though this NRC effort is not focused on causes of Arctic sea-ice loss (where black carbon has been identified as a possible contributor), further information about the implications of sea-ice loss could inform Arctic Council efforts to address climate pollutants.

SESSION #2
OBSERVATIONAL EVIDENCE OF LINKAGES

Arctic Connections to Extreme Weather: Evidence and Gaps
Jennifer Francis, Rutgers University

This presentation discusses evidence, and lack thereof, supporting the proposed chain of events that connects observed rapid warming of the Arctic with changing weather patterns in the mid-latitude northern hemisphere. Some links in this chain are well supported: 1) Arctic amplification (i.e., enhanced warming of the Arctic relative to the northern hemisphere as a whole), 2) a weakened poleward temperature (or atmospheric thickness) gradient as a result of the differential warming, 3) weakened upper-level zonal (west-to-east) winds in areas where poleward gradients have weakened, and 4) more meridional (i.e., (larger fraction in the north-south direction) upper-level winds in the same areas. The remaining three links in the chain are less solid, but the available evidence is generally supportive. More research, using both new analysis techniques of observations/model output as well as targeted modeling experiments, is clearly needed to either confirm or disprove the links. These include: 5) a more meridional upper-level flow leads to increased amplitude of large-scale (Rossby) waves and an increased likelihood of blocking, 6) a more amplified flow causes west-to-east wave propagation to slow, and 7) slower moving upper-level waves cause more persistent weather patterns, which increase the likelihood of extreme weather events associated with prolonged weather conditions. Additional research is also needed to untangle interactions among Arctic amplification and other large-scale circulation features, such as ENSO/PDO, AO/NAM/NAO, AMO, stratospheric circulation changes, etc.

The Link between Tropical Convection and the Arctic Warming on Intraseasonal and Interdecadal Time Scales

Steve Feldstein, Pennsylvania State University

During the past several decades, the largest signal of global warming in the atmosphere has been observed to occur over the Arctic Ocean. Coincident with this warming has been a large downward trend in Arctic sea ice. Several theories have been presented to explain this so-called Arctic Amplification. Here, evidenced is presented that the Tropically-Excited Arctic warMing (TEAM) mechanism makes an important contribution to this warming of the Arctic. The TEAM mechanism proposes that the trend toward stronger and more localized convection over the tropical Indo-Pacific Warm Pool excites poleward propagating Rossby wave trains which warm the Arctic by transporting heat and moisture poleward, inducing sinking motion/adiabatic warming over the Arctic, and increasing the downward flux of infra-red radiation.

Results are presented which show that Arctic warming is associated with particular atmospheric teleconnection patterns that fluctuate on a time scale of 5-10 days (the Pacific/North American and circumglobal teleconnection patterns). A link between the observed multi-decadal Arctic warming trend and these rapidly-evolving intra-seasonal time scale teleconnection patterns is shown to arise through an increase in the frequency of occurrence of these patterns over the past several decades.

An analysis of the impact of Arctic sea ice is also presented. It is first shown that Arctic sea-ice anomalies are associated with the dominant atmospheric teleconnections in the Northern Hemisphere. Furthermore, these sea-ice area anomalies are found to precede the teleconnection patterns with lead times of up to 12 months. This link between Arctic sea ice and the circulation over much of the Northern Hemisphere is found to occur through changes in the strength of the stratospheric polar vortex.

Observational Evidence of Arctic Weather Linkages

James E. Overland, NOAA/Pacific Marine Environmental Laboratory

Loss of Arctic sea ice, record negative values of the winter Arctic Oscillation atmospheric circulation index, earlier summer snow melt, and increasing extreme weather events at mid- latitudes—both heat waves and cold snowstorms—have been observed over the last decade. While the length of time series is too short to robustly differentiate Arctic forcing of mid-latitude extremes from random events, it is important to identify examples that support proposed linkages. Here, we focus on early winter and find three regionally different mechanisms operating in eastern North America, northern Europe, and far eastern Asia.

Similar penetrations of cold air into the eastern United States in December 2009, 2010, and 2012 relate to a shift in the long-wave upper-level atmospheric wind pattern. This shift coincides with warmer temperatures and greater geopotential thickness over northeastern Canada, major sea ice loss during October in Baffin Bay, a positive Greenland Blocking Index (greater 500 hPa geopotential heights), and record negative values of the Arctic Oscillation index. Such a combination of events amplified and shifted the climatological atmospheric wind pattern westward and allowed deeper southward penetration of cold air into the US. Northward air flow over Davis Strait acted as a positive feedback to maintain the higher air temperature anomalies. Anomalously cold early-winter weather in northern Europe has a direct teleconnection to loss of sea ice in the Barents and Kara Seas. Upper-level atmospheric circulation in north-central Asia in winter responds to the recent large scale reduction the north-south temperature gradients, as noted by Francis and Vavrus (2012), reducing jet-stream zonal velocities and the penetration of storms into northern Asia from the west, increasing the strength and persistence of the Siberian High, and thus increasing the intra-seasonal probability of multiple cold air events over eastern Asia. Mid-latitude attribution remains difficult and controversial as there are complex interactions of Arctic forcing with chaotic mid-latitude flow. One would not expect events to happen the same way every year even with similar Arctic forcing.

Linkages between Extratropical Weather and Arctic Weather

John R. Gyakum, McGill University

A. Apparent impacts of extratropical cyclogenesis on an extreme Arctic wind case (September 1999)

We study a case of an extreme storm surge, associated with a mesoscale surface cyclone, which occurred along the Beaufort Sea coastline during September 1999. Precursor meteorological conditions included an extreme water vapor transport into the Alaskan coast, and an extreme case of explosive cyclogenesis event affecting the southern Alaskan coast. A mesoscale modeling study has shown that latent heating associated with the explosive cyclone to the south was responsible for building a meridionally-oriented upper-tropospheric ridge that provides support for upshear surface cyclogenesis in the Beaufort Sea, whose flow contributes to the strong surface winds at Tuktoyaktuk. The local effects of the opening of the Beaufort Sea, though enhancing vulnerability, do not appear meteorologically important for such a storm surge event. If one component of future climate change is an eastward displacement of storm tracks into the northeastern Pacific, we may expect more such episodic events.

B. Apparent impacts of Arctic air mass formation on explosive extratropical cyclogenesis (January-February 1979)

The genesis of an arctic air mass in late January 1979 preceded a record cold-air outbreak in the United States. The associated planetary-scale flow facilitated a succession of synoptic-scale vorticity maxima that were responsible for a week-long clustering of explosive cyclogenesis. During this explosive cyclogenesis clustering, a strong increase in the North Atlantic basin's coverage of strong moist baroclinic growth rates occurred. Additionally, the largest Northern Hemispheric available energy peak was observed during the period just prior to the onset of the clusters of explosive cyclogenesis. This work shows that the process of arctic air mass generation in northern Alaska has significant planetary-scale consequences.

3-Minute Open Session Presentations: Observations

Open Panel Talk
Xubin Zeng, University of Arizona

Monthly diurnal temperature range (DTR) is simply the monthly average of daily (maximum – minimum) land surface air temperature difference. First, we find the zonally-averaged DTR in January over high latitudes to remain large, which is opposite to our physical understanding as diurnal solar forcing is zero. Further analysis of hourly temperature data indicates that these DTR values represent the synoptic movement of weather systems every few days (rather than the diurnal cycle). A more robust and appropriate metric (DTRh), computed as the amplitude of monthly-averaged hourly temperature diurnal cycle, is proposed to represent the diurnal cycle. Second, we find a negative DTR trend in January from 1979-2009 north of 40oN, consistent with previous studies. However, this does not represent the change of diurnal cycle with time. In fact, DTRh does not show any trend. These results suggest the revision of all previous studies on DTR over high latitudes in winter.

Differential Pole-Eq. Warming and Rainfall
Rit Carbone, National Center for Atmospheric Research

Slower tropospheric winds suggest weaker vertical shear of horizontal wind. Vertical shear of greater than or equal to 10-3 s -1 is a necessary condition for 50-60 percent of June-July-August (JJA) rainfall in the central United States. This phenomenon is common to many "breadbaskets" of the world. The NCAR Climate System Model (CSM) 4th Assessment runs from 2090s – 2000s present an "ill-posed finding" on vertical wind shear reduction in that because global climate system models cannot simulate the described phenomena, they likely have substantial systematic errors in winds, momentum fluxes and vertical shear. Therefore the significance of projected decreases in shear, while physically plausible and expected, is uncertain. This finding implies a major redistribution of continental warm season rainfall, both spatially and temporally within the diurnal cycle. This finding may be similarly applicable to major portions of China, South America, Australia, Europe, Africa and South/Southeast Asia. Convection-permitting global climate models are needed to resolve this issue.

Flash Floods and Drought: 1) An Indicator of Shifts in Downpours (Precipitation IDF Curves approach) and 2) 2012 Summer Anomalies and the U.S. Drought
Sam Higuchi, NASA

There are two approaches to examining the relationship between an increase in flash floods (Intensity-Duration-Frequency Curves) and Arctic amplification, as a leading indicator of change and as a lagging indicator of change. NOAA's traditional approach to lagging indicators of change assumes the stationary case, using traditional statistical approaches (based on the central limits theorem) results. At least one site-specific conclusion is based on computational results without looking at the graphed observational data. The non-traditional approach to leading indicators of change is splitting the observational data and graphing observational differences using Generalized Extreme Value theory. The statistical approach suggests non-stationary in some cases (there is a change in slope, similar in concept to "hinge-fit" of "new normals"). NOAA also examined the 2012 U.S. Drought and mid-latitude summer weather anomalies and the short-term weather signatures and long-term hints of climate shifts. The analysis was limited to North America, included no jet stream analysis, and no atmospheric blocking analysis. This should be looked at again based on new evidence.

Warm Arctic—Cold Continents, Common among Melting Sea Ice, Rapid Snow Advance and the Negative AO
Judah Cohen, Atmospheric and Environmental Research

Regressions between autumnal sea ice extent, Eurasian snow cover extent the Arctic Oscillation (AO) and Northern Hemisphere temperatures yield the characteristic 'warm Arctic – cold continents' pattern. This pattern was observed during winter 2012/13 and is found to be common among years with observed low fall sea ice, rapid fall Siberian snow cover advance and a negative winter AO. Dynamical models, however, fail to capture this pattern, showing instead maximum warming over the Arctic Ocean and widespread winter warming over the adjacent continents. Plotting the daily-standardized polar cap geopotential height from 10-1000 hPa (PCH) from October 2012 through March 2013 shows that episodic warming of the Arctic coincides with many extreme weather events across the mid-latitudes. Regression of September 2012 sea ice extent anomalies onto the PCH from October 2012 through March 2013, further suggests that the dramatic decrease in sea ice contributed to extreme weather events observed during that period.

Pacific Arctic Sea Ice Loss
Kathleen Crane, NOAA

Pacific Arctic sea ice loss from advection of heat from the Pacific and Atlantic Gateways is shown. Added ocean heat storage and heat flux from new sea ice free areas affects also the mid-latitudes. More observations are needed to improve boundary layer conditions in models (e.g., the thinning of ice from below using buoys). There are almost no observations in some key areas for heat transport. NOAA is working with Russian Academy of Sciences to improve these observations in the Arctic Ocean.

Probability Density Function of Standardized Anomalous Daily Sea Level Pressure–Based Index of the North Atlantic Oscillation
Matt Newman, University of Colorado

The probability density function (PDF) of standardized anomalous daily Sea-Level-Pressure based index of the North Atlantic Oscillation (NAO) is estimated for two 68-year periods, 1874 to 1942 and 1943 to 2010, both as raw Histograms and as fitted "SGS" PDFs. Results are shown in a figure for each one of the 56 members of the observational 20th century reanalysis ensemble. There are thus 56 curves in each plot for each period. The spread among the curves is a measure of observational uncertainty, whereas the grey swath is a measure of sampling uncertainty. The figure shown indicated that there is no statistically significant change in the PDF of the NAO from the first to the second 68-year period.

A Synoptic-Dynamic Analysis of the Intense Arctic Cyclone of Early August 2012
Lance Bosart, University at Albany/SUNY

Lance Bosart provided a brief overview of his presentation, "A Synoptic-Dynamic Analysis of the Intense Arctic Cyclone of Early August 2012," from the American Meteorological Society's 12th Conference on Polar Meteorology and Oceanography. The full abstract of that presentation is provided below:

A surface cyclone formed along an anomalously strong baroclinic zone over north-central Russia on 2-3 August 2012. This cyclone moved northeastward, intensified slowly, crossed the northeastern coast of Russia on 4 August, and strengthened rapidly as it moved poleward over the Arctic Ocean on 5-6 August. By 1200 UTC 6 August, this intense Arctic Ocean cyclone had achieved a minimum sea level pressure of < 965 hPa near 83 N and 170 W. This presentation is motivated by the likelihood that this cyclone was arguably the most intense storm system to impact the Arctic Ocean in the modern data record going back to the International Geophysical Year in 1957-1958. The purpose of this presentation will be to present the results of a synoptic-dynamic analysis of this intense early August cyclone to help gain a better understanding of why such intense cyclones are so rare over the Arctic Ocean.

Anticyclonic wave breaking in the upper troposphere across Russia in late July and very early August 2012 created an anomalously strong baroclinic zone between 60-80 N. Between 90 E and the Dateline, negative 850 hPa temperature anomalies between -2 and -4 C were found poleward of 70-75 N over the Arctic Ocean in the 1-5 August time mean. Likewise, positive time-mean 850 hPa temperature anomalies upwards of 8-9 C were situated over eastern Russia near 60 N. In response to this observed temperature anomaly pattern, an anomalously strong 850 hPa temperature gradient of ~10 C (2000 km)-1 between 60-80 N helped to sustain an anomalously strong (20-25 m s-1) 250 hPa jet along the coast of northeastern Russia. A local wind speed maxima along this 250 hPa jet corridor reached 40-50 m s-1 immediately upstream of the surface cyclone.

Because the surface cyclone intensified most rapidly over the relatively ice free Arctic Ocean in the poleward exit region of the aforementioned jet streak, the question arises as to how much sensible and latent heat fluxes from the relatively ice free Arctic Ocean contributed to destabilizing the lower troposphere and augmenting the dynamically driven component of the observed cyclogenesis. Likewise, unusually high observed 1000-500 hPa thickness values between 564-570 dam in the warm sector of the developing cyclone over north-central Russia were indicative of the strength of the cyclone warm sector and the ability of warm-air advection to sustain deep ascent. We will attempt to distinguish the relative importance of dynamical versus thermodynamical forcing to the cyclogenesis process in our presentation.

Declining Spring Snow Cover Extent over High Latitude Northern Hemisphere Lands
David A. Robinson, Rutgers University

Annual snow cover extent (SCE) over Northern Hemisphere (NH) lands has averaged lower since the late 1980s than earlier in the satellite era that began in the late 1960s. This is most evident from late winter through spring, and in the past decade has been exceedingly pronounced at high latitudes in May and June. Monthly SCE is calculated at the Rutgers Global Snow Lab from daily SCE maps produced by meteorologists at the National Ice Center. The most recent four Mays have had four of the five lowest NH SCEs on record, with Eurasian (Eur) SCE at a record low in 2013. North American (NA) SCE achieved a record minimum in May 2010, but of late has not been as consistently low as over Eurasia. The past six Junes have seen record minimum SCEs over NH and Eur, with five of these six the lowest over NA. The recent early timing of arctic snowmelt appears to be occurring at an equivalent if not greater pace than the loss of summer Arctic sea ice extent.

SESSION #3
THEORETICAL AND MODELING WORK

Atmospheric Energy Transports, Polar Amplification, and Mid-latitude Climate
Dargan M. W. Frierson, University of Washington

A variety of general circulation model studies in recent years have shown that changes in high latitudes affect climate even in faraway regions of the globe. Since many of the mechanisms for such connections are based on energy transports, we review expectations for how the poleward transport of energy into the Arctic might change in the future. In simulations of global warming, increased latent energy transport is compensated by reduced dry static energy transport. The total atmospheric transport is actually *anticorrelated* with polar amplification across models, and some of the models with the most polar amplification have anomalous energy transport *towards* the mid-latitudes. We suggest that the response of energy transports to Arctic amplification is first order diffusive, meaning some of the high latitude warming will be spread into mid-latitudes via reduced dry static energy transport.

However, there still can be some local mid-latitude cooling and enhanced extremes in response to rapid Arctic amplification. We propose a new chain of dynamical links that could connect Arctic amplification to mid-latitude weather extremes: decreased temperature gradients and modified high latitude wind profiles should induce a negative phase of prominent oscillation indices such as the NAO, and we suggest that either baroclinic or barotropic mechanisms could cause such a shift. A negative phase NAO index is associated with regions of local cooling, more Greenland blocking, and more temperature extremes. These dynamical links should be tested in model simulations, and regional and seasonal sensitivities should also be evaluated.

Response of the Wintertime Atmospheric Circulation to Current and Projected Arctic Sea Ice Decline

Gudrun Magnusdottir, University of California, Irvine

Motivated by the rapid September decrease in sea-ice concentrations over the years 2007-2012 (inclusive), Yannick Peings and I recently planned and ran some idealized CAM5 experiments using sea-ice forcing from those 6 years (compared to average sea ice conditions over 1979-2000). Our motivation was in part to isolate effects from these few years of rapid decline in sea-ice forcing from other possible forcing mechanisms in the Arctic that would be present in the observational record. Our secondary motivation was to compare results from this experiment (called 2010C) to another experiment where we forced the model with projected sea-ice concentrations corresponding to 2080-2099 (called 2090C).

The surprising result is that the 2010C experiment (or current conditions scenario) forces a remote atmospheric response in late winter that favors cold land surface temperature over mid-latitudes as has been observed in recent years. Anomalous Rossby waves forced by the sea ice anomalies penetrate into the stratosphere in February and weaken the stratospheric polar vortex, resulting in negative anomalies of the Northern Annular Mode (NAM) that propagate downwards during the following weeks, especially over the North Pacific. The seasonality of the response is attributed to the timing of the phasing between the anomalous and climatological waves. When sea ice concentration taken from projections of conditions at the end of the 21st century is prescribed to the model, negative anomalies of the NAM are visible in the troposphere, both in early and late winter. Little impact is found in the stratosphere in this experiment, and this NAM-signature is driven by the large warming of the lower troposphere over the Arctic. A weak but significant increase of the mid-latitude meanders is identified, however, the thermodynamical response extends beyond the Arctic and offsets the dynamical effect at least north of 45 N. Thus, the stronger sea ice forcing results in no stronger intensity of cold extremes over mid-latitudes.

This latest version of the NCAR AGCM (CAM5) has been improved in terms of representing the seasonal cycle of Arctic clouds (Kay et al 2012, J Clim, 5190-5207). Perhaps more importantly the stratospheric circulation is truer to observations (was too strong in previous versions of the model). We haven't really addressed the reasons for the remote response now whereas none was found in older versions of the model.

Arctic Sea Ice Predictability and its Long-term Loss and Implications for Ocean Conditions
Marika Holland, National Center for Atmospheric Research

Climate models have been very useful tools for exploring linkages between Arctic sea ice loss and climate. In particular, the loss of Arctic sea ice has implications for ocean circulation in the north Atlantic, with resulting effects on poleward ocean heat transport. The magnitude of these simulated changes is strongly related to the magnitude of ice loss (which is nearly linear with the atmospheric CO_2 increase). Using Arctic sea ice predictions to forecast future weather has real limitations because our ability to predict the sea ice itself is so limited.

The seasonal prediction of Arctic sea ice is an initial value problem. With almost perfect knowledge of initial conditions, there are inherent limits on predictability. One can examine how ensemble members diverge over time and compare that to the natural variability of the system; when these are indistinguishable, predictability is lost. The model studies examined in this presentation use simulations from CCSM3. Similar initial conditions were used to do an ensemble of predictions with slightly varying initial atmospheric conditions in each run. Results show that the potential prognostic predictability (PPP) decreases during spring and is regained during summer and following winter. There is significant winter predictability due to memory in ocean heat content. The memory of ice edge location is associated with sea surface temperature (SST) predictability. The model studies suggest limited predictability in sea ice area on 1-2 years timescales, particularly in winter. It is currently unclear what spatial coverage, variables, etc. are needed for "adequate" observational network, as well as what model biases impact predictability and what model complexity is required.

The current opportunities for using Arctic sea ice predictions to assess the risk of temperature and precipitation anomalies and extreme weather events over northern continents are subject to inherent limits of the sea ice prediction system. Sources of uncertainty in projections of future climate change include intrinsic climate variability, and the fact that climate models simulate the statistics of climate and not the events of any particular year. North Atlantic circulation changes lead to reduced ocean heat transport and reduced warming. Regarding the impact of longer-term ice-loss changes on mid-latitude weather, these ocean changes need to be considered. Much of the mid-21st century North Atlantic (and high latitude) surface change can be attributed to summer sea ice loss.

3-Minute Open Session Presentations: Modeling

The Role of Global Climate Change in the Extreme Low Summer Arctic Sea Ice Extent in 2012
Rong Zhang and Thomas R. Knutson, NOAA/GFDL

Comparisons between observations and 19 CMIP5 models reveal that the 2012 summer Arctic sea ice extent (ASIE) anomaly and the rapid decline of summer ASIE in the early 21st century are very rare occurrences in the context of these models and their responses to anthropogenic and natural forcing combined. The observed 2012 record low in summer ASIE is extremely unlikely to have occurred due to internal climate variability alone, according to the models, and has a much greater likelihood of occurrence in the "forced plus internal variability" scenario. The observed September ASIE decline trend for 2001-2012 is much more rapid than in the previous two decades and even lies outside of the 5th-95th percentile range of the multi-model distribution of forced responses, despite the observed September global mean surface air temperature (SAT_{gm}) warming trend for the same period being smaller than in previous decades.

The Influence of Sea Ice Albedo on the Global Hydrological Cycle

Aneesh Subramanian, Ian Eisenman, and Simona Bordoni, Scripps Institution of Oceanography, University of California, SanDiego

The idea that Arctic sea ice retreat could influence precipitation outside the polar region has garnered considerable interest in recent years. As sea ice recedes, it alters the climate system due to factors including a change in surface albedo, which has been suggested to affect the hydrological cycle in the mid-latitudes and tropics. Here we examine how sea ice influences the response of the global hydrological cycle to climate change in the broader context of climates ranging from an ice-covered earth to an ice-free earth. We use an idealized general circulation model with annual-mean forcing and the surface albedo set to an ice-covered value where the temperature is below the freezing point and an ice-free value elsewhere. We assess the importance of sea ice albedo by comparing the simulation that includes ice albedo effects with a second simulation that has the same global-mean surface temperature but has the albedo set to the ice-free value everywhere (even where ocean temperatures fall below the freezing point). This work builds on a body of previous theories developed in idealized frameworks that did not include surface albedo changes. We find that ice albedo effects cause substantial changes to the global hydrological cycle in cold climates, but not in warm climates, with the transition occurring at a point relatively near to the present-day climate. Intriguingly, we find that the inclusion of ice albedo effects in cold climates causes an increase in global-mean precipitation despite a decrease in absorbed shortwave radiation and hence less energy available to evaporate water. Large-scale condensation, rather than convection, is the main contributor to the changes in climates close to present day. We explain this result with the finding that ice albedo effects cause a steeper temperature gradient and hence shorter distances for air parcels to travel along isentropes before reaching saturation (a thermodynamic effect) as well as a more vigorous atmospheric circulation (a dynamic effect). These results suggest that ice albedo effects are important for understanding the hydrological cycle in climates colder than today, including many paleoclimate environments.

Arctic Influence on Weather and Climate in Japan

Jinro Ukita, Niigata University

At present there remain unclear as to both mechanisms and implications of linkages between Arctic and mid-latitudes changes in weather and climate. In Japan we have established a research project specifically addressing this potentially significant climate question (http://www.nipr.ac.jp/grene/e/index.html). We are currently investigating possible Arctic impacts on NH weather and climate using a JAMSTEC-AGCM with the model configuration of T79L56 and the model top set at 0.08hPa. The numerical experiment consists of two perpetual runs with prescribed SST and sea ice concentration (SIC) conditions from the 1979-1983 and 2005-2009 periods, respectively. In the case where the only difference in the boundary condition is from NH SIC there appears statistically significant cooling in the eastern Siberia and weaker cooling in the eastern Europe and northeastern North America in the temperature difference field. Further analysis reveals that this is due to southward propagation of the Rossby waves aloft originated from an anomalous heat source over the Arctic.

Open Session Talk
Uma Bhatt, University of Alaska, Fairbanks

Daily sea ice concentration tendency in five CMIP5 simulations is compared with observations to reveal that most models underestimate this quantity that describes high-frequency ice movements, particularly in the marginal ice zone. Does underestimating this high-frequency ice variability matter to the atmosphere? A set of global climate model simulations were conducted with prescribed sea ice and demonstrate that the atmosphere responds differently when daily ice variations are included. Sea ice differences in September lead to an anomalous high and weaker storm activity over northern Europe. During October, the Arctic ice expands equator-ward faster with daily ice and leads to a local response of near-surface cooling. In DJF, there is a 1.5-hPa positive sea level pressure anomaly over North America, leading to anomalous northerly flow and anomalously cool continental U.S. temperatures. The differences arising from high temporal frequency ice variability, while modest, are relevant for sea ice impacts on mid-latitude weather.

Is the recent sea ice trend a Rapid Ice Loss Event (RILE)?
Cecilia Bitz, University of Washington

RILEs are periods of extreme loss of sea ice extent in a five-year or longer period. We assess the probability of RILEs in the CMIP5 models for 1850-2100 (RCP8.5 scenario) in 84 ensemble members from 25 models. RILEs occur on average 3 times in each ensemble member. Most RILES occur during the first half of the 21st century. The probability of a RILE in a given five year period is about 10% at present, increasing to about 20% by 2040. Before 2040, RILES are increasingly common as the sea ice thins because the radiative forcing is also increasing during the 21st century in the RCP8.5 scenario and because the natural variability of the sea ice is also increasing. RILEs become quite rare in the late 21st century because many models are essentially ice-free at that time. About 5 years after a RILE, there is no lasting impact and the sea ice extent returns roughly to a background rate of loss that is similar to the loss rate prior to the RILE. Hence, there is no evidence of tipping point behavior in the CMIP5 models.

Our analysis indicates that it is possible we have experienced a RILE this decade, and possibly now we are on a trajectory back towards the background rate of ice loss. Our analysis indicates that at a given time, it is impossible to distinguish a RILE from a continuously accelerating rate of loss because the difference depends on the future: A RILE necessarily ends with a period of weaker sea ice loss, while a continuously accelerating loss has increasingly greater loss with each year.

Open Session Talk
Clara Deser, NCAR

We investigated the role of oceanic feedbacks in the atmospheric response to GHG-induced Arctic sea ice loss. To do this, we performed atmosphere-only and coupled atmosphere-ocean model integrations with the same late 21st century Arctic sea ice conditions specified as a lower boundary condition. The future sea ice conditions were taken from a fully-coupled model integration forced with the RCP8.5 GHG scenario. In this way, we were able to isolate the role of ocean coupling in the atmospheric response to Arctic sea ice loss. The results show that the primary effect of having an interactive ocean is to deepen the atmospheric temperature response to the mid-troposphere (about 400 hPa) compared to the atmosphere-only run in which the thermal response is confined to the planetary boundary layer (below 850 hPa). This result has implications for the role of high-latitude ocean feedbacks in Arctic Amplification. The near-surface atmospheric circulation response did not appear to be sensitive to ocean coupling.

Appendix B
Workshop Agenda and Participant List

Board on Atmospheric Sciences and Climate
and
Polar Research Board

Linkages Between Arctic Sea Ice Loss and Mid-Latitude Weather Patterns

September 12-13, 2013

Earth System Science Interdisciplinary Center
University of Maryland
5825 University of Maryland Research Park
College Park, MD 20740

Workshop Goals

Rising global average temperatures, and especially intense warming in the northern polar regions, are leading to a rapid loss of the sea ice cap that covers the Arctic ocean. Emerging research may indicate that large losses of Arctic sea ice cover can have dramatic impacts upon weather patterns across the heavily populated northern mid-latitudes, and that such impacts could increase as ice cover continues to retreat in the coming decades. The workshop will address the following questions:

- What do we currently understand about the mechanisms that link declines in Arctic sea ice cover, loss of high-latitude snow cover, changes in Arctic -region energy fluxes, atmospheric circulation patterns, and the occurrence of extreme weather events?
- What may be the possible implications of more severe loss (and eventually, total loss) of summer Arctic sea ice upon weather patterns at lower latitudes?
- What are the major gaps in our understanding, and what sort of observational and/or modeling efforts are needed to fill those gaps?
- What are the current opportunities and limitations for using Arctic sea ice predictions to assess the risk of temperature/precipitation anomalies and extreme weather events over northern continents? How might these capabilities improve over time?

Thursday, September 12, 2013

OPEN SESSION: 9:00 A.M.–5:30 P.M.

9:00 A.M. Welcome, Introduction, Purpose of Workshop — David Robinson, Rutgers University

Session 1

Big picture context: The role of Arctic sea ice loss vs. other forcing factors

9:15 A.M. James Screen, University of Exeter
Elizabeth Barnes, Colorado State University
Paul Kushner, University of Toronto

12:00 P.M. *Lunch*

1:00 P.M. Open panel session for workshop participants

Session 2

Observational evidence of linkages

1:30 P.M. Jennifer Francis, Rutgers University
Steve Feldstein, Pennsylvania State University
James Overland, NOAA
John Gyakum, McGill University

5:00 P.M. Open panel session for workshop participants

5:30 P.M. Adjourn

Friday, September 13, 2013

OPEN SESSION: 8:30 A.M.–3:00 P.M.

Session 3

Theoretical and modeling work

8:30 A.M. Dargan Frierson, University of Washington
Gudrun Magnusdottir, University of California, Irvine
Marika Holland, NCAR

11:15 A.M. Open panel session for workshop participants

12:00 P.M. *Working lunch in breakout groups*

Session 4

Breakout group discussions

Where do we go from here? What are the major gaps in our understanding, and what observational and modeling efforts are needed to fill those gaps?

1:45 P.M. Breakout groups present their findings

3:00 P.M. Workshop adjourns

Linkages Between Arctic Sea Ice Loss and Mid-Latitude Weather Patterns

September 12-13, 2013

Participant List

- Anjuli Bamzai, NSF
- Elizabeth (Libby) Barnes, Colorado State University
- Uma Bhatt, University of Alaska, Fairbanks
- Cecilia Bitz, University of Washington
- Lance Bosart, University at Albany/SUNY
- David Bromwich, Ohio State University
- Tony Busalacchi, University of Maryland
- Richard (Rit) Carbone, NCAR
- Jessie Carman, NOAA
- Judah Cohen, Atmospheric and Environmental Research (AER)
- Kathleen Crane, NOAA
- Ed Dunlea, NRC
- Benjamin DeAngelo, EPA
- Clara Deser, NCAR
- Ian Eisenman, University of California, San Diego
- Daniel Eleuterio, US Navy
- Steve Feldstein, Pennsylvania State University
- Jennifer Francis, Rutgers University
- Dargan Frierson, University of Washington
- Rita Gaskins, NRC
- Laurie Geller, NRC
- Rob Greenway, NRC
- Michael Gremillion, US Air Force
- John Gyakum, McGill University
- Greg Hakim, University of Washington
- Wayne Higgins, NOAA
- Sam Higuchi, NASA
- Martin Hoerling, NOAA
- Marika Holland, NCAR
- Fiona Horsfall, NWS
- Paul Houser, George Mason University
- Brendan Kelly, OSTP
- Paul Kushner, University of Toronto
- Michelle L'Heureux, NOAA
- Arthur Lee, Chevron
- David Lorenz, University of Wisconsin
- Gudrun Magnusdottir, University of California, Irvine
- Walter Meier, NASA
- Matt Newman, University of Colorado
- J. Jerome Montague, Alaskan Command
- Jim Overland, NOAA
- Amanda Purcell, NRC
- Susan Roberts, NRC
- David Robinson, Rutgers University
- James Screen, University of Exeter
- Amanda Staudt, NWF
- Julienne Stroeve, NSIDC
- Katie Thomas, NRC

- Steve Tracton, Washington Post
- Jinro Ukita, Niigata University
- Stephen Vavrus, University of Wisconsin
- Tom Wagner, NASA
- John Walsh, University of Alaska, Fairbanks
- Wanqiu Wang, NOAA
- Tim Woollings, Reading University
- Xubin Zeng, University of Arizona
- Rong Zhang, NOAA

Appendix C
Table and Schematic Presented by Session 1 Speakers

Session 1 speakers presented the table below as a potential tool for organizing information about what is known and not known about proposed linkages between Arctic warming and mid-latitude weather patterns. The first column is used to describe each linkage. The second column is meant to assess confidence in and strength of linkages by outlining our current scientific understanding of that linkage. The ACID test (Attributable, Corroborated, Informed, Detected) could be incorporated here. The third column is used to identify critical research gaps in observations and models. The first row provides some suggestions on factors to consider. The remaining two rows provide examples.

The schematic is illustrative of the big picture questions beyond the scope of the workshop:

- To what does the Arctic respond?
    - Changes at the Polar Cap boundary.
    - Changes at other boundaries (surface, tropopause, composition of atmospheric column).
- To what does the mid-latitude circulation respond?
    - Changes at the Polar Cap boundary.
    - Changes at other boundaries (surface, tropopause, subtropical boundary, composition).

Arctic Linkages: Summary table		
a. Linkage and Description: • Timescale (intraseasonal, decadal, …) • Season (JFM, SON, …) • Proposed role of Arctic (direct, through feedbacks, through modes) • Motivation to study (fundamental, forecasting, socioeconomic, ecological)	**b. Current Understanding:** • Basis of evidence (theory, obs, models) • Related linkages, and separability from these linkages • Level of confidence (ACID) • Quantification (e.g. sensitivity factors *M*/ *A*)	**c. Research Gaps:** • Observational capacity • New observational analysis • Modeling capacity • New model experiments and diagnostics • Theoretical knowledge • Impacts analysis
1a. Linkage and Description: **Sea ice and mid-latitude wave amplitude events** • Multidecadal trend • Seasonality of response is poorly understood • Proposed to be forced by Arctic amplification; direct Arctic involvement • Connected to mid-latitude weather extremes	**1b. Current Understanding:** • Observational evidence: some in support, some not; metric dependent • Some theoretical arguments in support, but details of dynamical mechanism are lacking. • Related to other implications of polar amplification. • Fundamental problem of signal to noise in analyses • Dynamical mechanism poorly quantified (e.g. seasonality of forcing and response, timescales, amplitude of response) • **A**ttributable – No; **C**orroborated – No; **I**nformed by mechanisms – Maybe; **D**etected –No	**1c. Research Gaps:** • Broader observational analysis required involving statistics of extremes. • Simplified model experiments possibly useful. • Studies of CMIP5 require high frequency output. • Long model runs, multiple realizations, required.
2a. Linkage and Description: **Eurasian snow/sea ice linkage** • Interannual variability and decadal trends • September sea ice and October snow • Arctic role: direct, through moisture transport	**2b. Current Understanding:** • Difficult to detect clear linkage apart from trends. • Disagreement with trends for different observational datasets • Model support: direct sea ice perturbation experiments • Physically plausible. • ACID assessment: **A**ttributable – in models; **C**orroborated – Only partially; **I**nformed by mechanisms – Partially; **D**etected – Partially.	**2c. Research Gaps:** • Need to consolidate climate data records • Need to analyze trends in models and understand connections to moisture transport.

How can we improve this schematic?

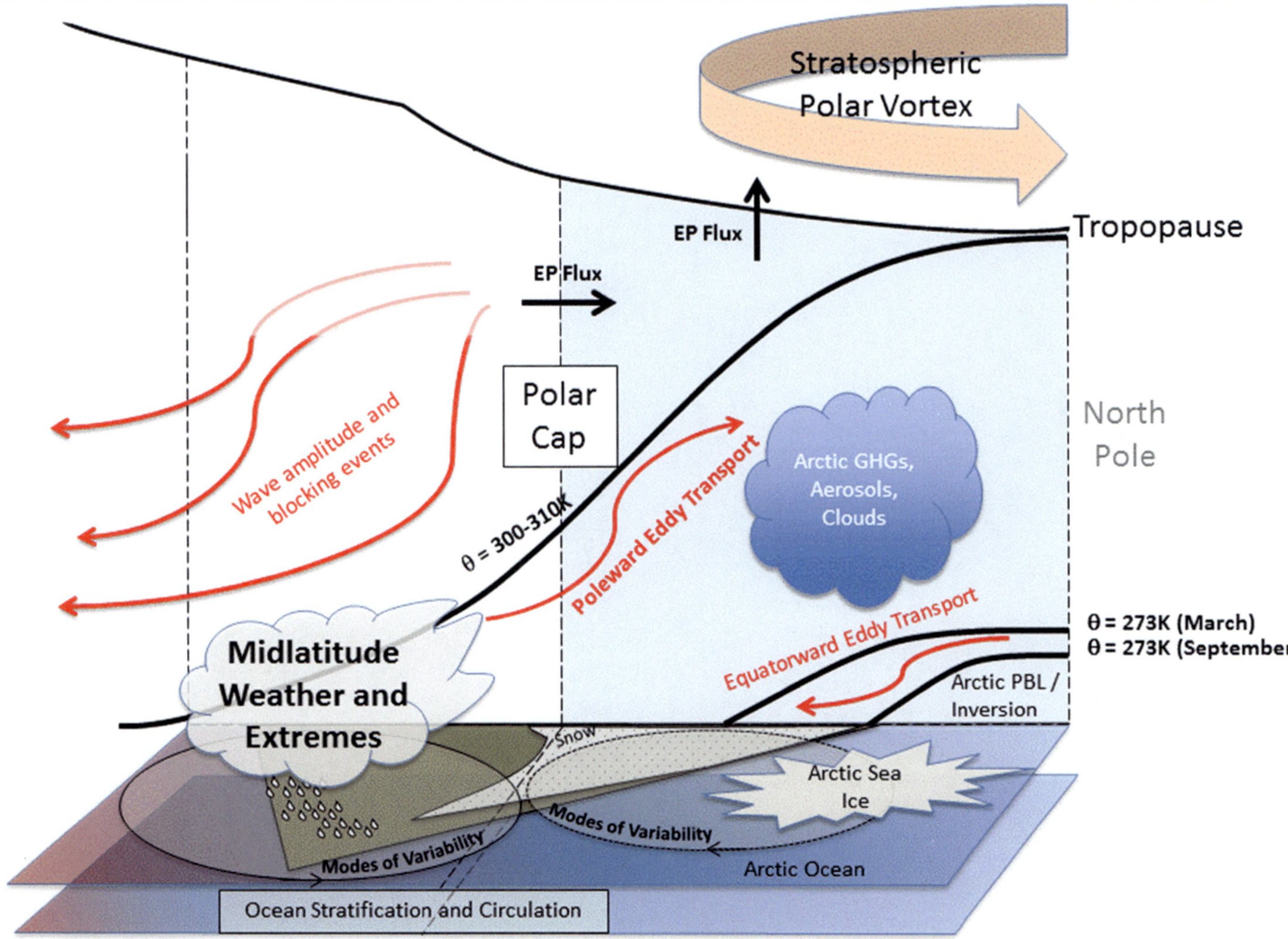

Appendix D
Acronyms and Initialisms

ACID	Attribution; Corroboration; Informed by mechanistic understanding; Detection
AMJ	April May June
AMOC	Atlantic meridional overturning circulation
CAM5	Community Atmosphere Model 5
DJF	December January February
EA	East Atlantic
ENSO	El Niño/Southern Oscillation
ESRL/PSD	Earth System Research Laboratory/Physical Sciences Division
GHG	Greenhouse Gas
JAS	July August September
JFM	January February March
JJA	June July August
MAM	March April May
NAM/AO	Northern Annular Mode/Arctic Oscillation
NAO	North Atlantic Oscillation
NCAR	National Center for Atmospheric Research
NCEP	National Centers for Environmental Prediction
NH	Northern Hemisphere
NIC	National Ice Center
NOAA	National Oceanic and Atmospheric Administration
NSIDC	National Snow and Ice Data Center
OND	October November December
PDO	Pacific Decadal Oscillation
RCP	Representative Concentration Pathway
SON	September October November
TEAM	Tropically-Excited Arctic warMing
VIRS	Visible and Infrared Scanner

Appendix E
Biographical Sketches of Planning Committee Members

Dr. David A. Robinson
Rutgers, The State University of New Jersey

Dr. David A. Robinson is a Professor at Rutgers, The State University of New Jersey. He has expertise in the collection and archiving of accurate climatic data and is interested in climate change (particularly state and regional climate issues), hemispheric and regional snow cover dynamics, interactions of snow cover with other climate elements, and the dynamics of solar and terrestrial radiative fluxes at and close to the surface of the earth. Dr. Robinson is the author or co-author of approximately 130 articles, more than half in peer-reviewed journals and book chapters. He also is the State Climatologist for New Jersey. Dr. Robinson has served on several NRC committees and as the chairman of the Committee on Climate Data Records from Operational Satellites: Development of a NOAA Satellite Data Utilization Plan. He received his Ph.D. from Columbia University.

Dr. Uma Bhatt
University of Alaska, Fairbanks

Dr. Uma Bhatt is an Associate Professor of Atmospheric Sciences at the University of Alaska, Fairbanks. Her research interests include role of sea ice, oceans, and land in climate variability, and change; multi-decadal climate variability; climate in Alaska; and complex systems research. She received her Ph.D. in atmospheric sciences from the University of Wisconsin—Madison in 1996.

Dr. Cecilia Bitz
University of Washington

Dr. Cecilia Bitz is an Associate Professor in the Atmospheric Sciences Department at the University of Washington. Her research interests include climate dynamics, climate change, paleoclimate, the role of sea ice in the climate system, Arctic/North Atlantic interactions, and sea ice model development. The primary tools for her research are a variety of models, from simple reduced models to sophisticated climate system models. Dr. Bitz is a member of the Advisory Board for the Community Climate System Model sponsored by the National Science Foundation and Department of Energy and the steering committee for the National Oceanic and Atmospheric Administration (NOAA) Climate and Global Change Postdoctoral Program. She received her Ph.D. in atmospheric sciences from the University of Washington in 1997.

Dr. David H. Bromwich
The Ohio State University

Dr. David H. Bromwich is a Senior Research Scientist and Director of the Polar Meteorology Group at the Byrd Polar Research Center of The Ohio State University. He is also a Professor with the Atmospheric Sciences Program of the Department of Geography. Dr. Bromwich's research interests include the climatic impacts of the Greenland and Antarctic ice sheets; coupled mesoscale-global circulation model simulations; the atmospheric moisture budget of high southern latitudes, Greenland, and the Arctic basin using numerical analyses; and the influence of tropical ocean-atmosphere variability on the polar regions. Dr. Bromwich has served on the National Research Council's Committee on Geophysical and Environmental Data and was previously a U.S. Representative of the Scientific Committee on Antarctic Research. He is a member of the American Meteorological Society, the American Geophysical Union, the Royal Meteorological Society, and the American Association of Geographers. Dr. Bromwich earned his Ph.D. in meteorology from the University of Wisconsin—Madison in 1979.

Dr. Lance F. Bosart
State University of New York, Albany

Dr. Lance F. Bosart is a Distinguished Professor in the Department of Earth and Atmospheric Sciences at the State University of New York, Albany. He joined the University at Albany faculty after he received his Ph.D. in meteorology from the Massachusetts Institute of Technology in 1969. He was promoted to full Professor in 1983 and Distinguished Professor in 2004. His research specialty is synoptic-dynamic meteorology. Dr. Bosart works on a variety of observationally driven large-scale, synoptic-scale, and mesoscale basic research problems that focus on gaining a better understanding of the behavior of tropical, mid-latitude, and polar weather systems.

Dr. Clara Deser
National Center for Atmospheric Research

Dr. Clara Deser is a Senior Scientist and Head of the Climate Analysis Section in the Climate and Global Dynamics Division at the National Center for Atmospheric Research. Her research interests include diagnostic analysis of observed and modeled global climate variability in the coupled atmosphere-ocean-ice-land system on diurnal to centennial time scales. Dr. Deser has worked in the field of climate variability throughout her career. Recent scientific activities have focused on the causes and consequences of Arctic sea ice loss, uncertainty in climate change projections, the El Niño Southern Oscillation phenomenon, the North Atlantic Oscillation, and the Pacific Decadal Oscillation. She received her Ph.D. in atmospheric sciences from the University of Washington in 1989.

Dr. Walter Meier
NASA

Dr. Walter Meier is a Research Scientist at NASA Goddard Space Flight Center. He was formerly a Research Scientist for the National Snow and Ice Data Center at the University of Colorado, Boulder. He is an expert on sea ice remote sensing and data assimilation and Arctic climate and climate change. His areas of observational expertise include SSM/I passive microwave polar stereographic sea ice products; visible and infrared products; and field observations. Dr. Meier is currently focused on better understanding the decreasing Arctic summer sea ice cover and its impacts. He received his Ph.D. from the University of Colorado, Boulder in 1998.